Contents

24-th Spanish Olympiad ◆ Hungarian National Olympiad 1987 ❑ 24-th Spanish Olympiad
Romanian Mathematical Olympiad, Final Round 1978 ○ Kozepiskolai Matematikai Lapok 1983
Asian Pacific Mathematical Olympiad 1989 ✧ IMO Proposal by Greece (1989) ⋏
Australian Mathematical Olympiad 1985 ○ 11th Austrian-Polish Mathematics Competition ❑
Singapore MSI Mathematical Competition 1988 ◆ First Selection Test of the Chinese I.M.O. Team
43rd Mathematical Olympiad (1991-92) in Poland (Final round) ❑ Canadian MO 1996 ○
10th Iranian Mathematical Olympiad (Second Stage Exam) ○ 6th Irish Mathematical Olympiad
6th Korean Mathematical Olympiad (Final Round), 1993 ⋏ Turkish Mathematical Olympiad 1993
Canadian Mathematical Olympiad 1997 ◆ 16th Austrian Polish Mathematical Competition ✧
Romanian First Team Selection Test 1993 (34th IMO) ○ **Mathematical Contest Baltic Way 1992**
Czechoslovakia Mathematical Olympiad 1993 ❑ Israel Mathematical Olympiad 1994
3rd Mathematical Olympiad of the Republic of China 1994 ✧ Swedish Mathematical Olympiad
7th International Mathematical Olympiad 1996 (Shortlist) ◆ Balkan Mathematical Olympiad 94
Croatian National Mathematics Competition 1994 ○ Kozepiskolai Matematikai Lapok 1983
17th Austrian-Polish Mathematics Competition 1994 ✧ IMO Proposal by Greece (1989) ⋏
38th Mathematics Competition of the Republic of Slovenia ❑ Israel Mathematical Olympiad 95
45th Latvian Mathematical Olympiad 1994 ❑ 18th Austrian-Polish Mathematics Competition 95
4th Mathematical Olympiad of the Republic of China (Taiwan) ○ Vietnamese Math Olympiad 96
31st Spanish Mathematical Olympiad 1994 ⋏ 3rd Turkish Mathematical Olympiad 1995 ○
Australian Mathematical Olympiad 1996 ◆ 10th Nordic Mathematical Contest 1996 ✧
9th Irish Mathematical Olympiad 1996 ○ **St. Petersburg City Mathematical Olympiad** 1996
Republic of Moldova XL Mathematical Olympiad 1996 ❑ Ukrainian Mathematical Olympiad 96
Taiwan Mathematical Olympiad 1996 ✧ Croatian Mathematical Olympiad 1995 ⋏
13th Iranian Mathematical Olympiad 1995 ❑ 6th ROC Taiwan Mathematical Olympiad 1997
Estonian Mathematical Contest 1995 (Final Round) ○ XXXIII Spanish Mathematical Olympiad 96
20th Austrian-Polish Mathematical Competition 1997 ❑ Chinese Mathematical Olympiad 1997
Swedish Mathematical Competition 1996 ⋏ Math Olympiad in Bosnia and Herzegovina 1997
Fourth National Mathematical Olympiad of Turkey 1997 ✧ Ukrainian Mathematical Olympiad 97
36th Armenian National Mathematical Olympiad 1997 ◆ Vietnamese Team Selection Test 1997
28th Austrian Mathematical Olympiad 1997 ❑ Iranian Mathematical Olympiad 1997
Swedish Mathematical Competition 1997 ○ Ukrainian Mathematical Olympiad 1998
1st Mediterranean Mathematical Olympiad 98 ⋏ Hungary-Israel Mathematical Competition 99
12th Korean Mathematical Olympiad 1999 ✧ Grosman Memorial Mathematical Olympiad 1999
Russian Mathematical Olympiad 1999 ◆ 16th Iranian Mathematical Olympiad 1999
Vietnamese Mathematical Olympiad 1999 ❑ Turkish Team Selection Test for 40th IMO 1999
Japanese Mathematical Olympiad 1999 ○ XLIII Mathematical Olympiad of Moldova 1999 ....

f(u)nctions and p(o)lynomials
problems and solutions
from
Mathematical Olympiads

24-th Spanish Olympiad - First Round (5)

Given the function f defined by $f(x) = \sqrt{4 + \sqrt{16x^2 - 8x^3 + x^4}}$.
 (a) Draw the graph of the curve $y = f(x)$.
 (b) Find, without the use of integral calculus, the area of the region bounded by the straight lines $x = 0, x = 6, y = 0$ and by the curve $y = f(x)$. Note: all the square roots are non-negative.

24-th Spanish Olympiad - First Round (7)

Let $I_n = (n\pi - \pi/2, n\pi + \pi/2)$ and let f be the function defined by $f(x) = \tan x - x$.
 (a) Show that the equation $f(x) = 0$ has only one root in each interval I_n, $n = 1, 2, 3, \ldots$.
 (b) If c_n is the root of $f(x) = 0$ in I_n, find $\lim_{n \to \infty} (c_n - n\pi)$.

24-th Spanish Olympiad (2)

Let f be a continuous function on $\mathbf{R}$ such that
 (i) $f(n) = 0$ for every integer n, and
 (ii) if $f(a) = 0$ and $f(b) = 0$ then $f(\frac{a+b}{2}) = 0$, with $a \neq b$.
Show that $f(x) = 0$ for all real x.

Hungarian National Olympiad 1987 (3)

Determine the minimum of the function
$$f(x) = \sqrt{a^2 + x^2} + \sqrt{(b-x)^2 + c^2}$$
where a, b, c are positive numbers.

Hungarian National Olympiad 1987 (11)

The domain of function f is $[0, 1]$ and for any $x_1 \neq x_2$
$$|f(x_1) - f(x_2)| < |x_1 - x_2|.$$
Moreover, $f(0) = f(1) = 0$. Prove that for any x_1, x_2 in $[0, 1]$,
$$|f(x_1) - f(x_2)| < \frac{1}{2}.$$

Romanian Mathematical Olympiad, Final Round 1978 (2)

Let P and Q be two polynomials (neither identically zero) with complex coefficients. Show that P and Q have the same roots (with the same multiplicities) if and only if the function $f : \mathbf{C} \to \mathbf{R}$ defined by $f(z) = |P(z)| - |Q(z)|$ has a constant sign for all $z \in C$ if it is not identically zero.

Kozepiskolai Matematikai Lapok 1983 (375)

Does there exist a function $f : \mathbf{R} \to \mathbf{R}$ such that $\lim_{x\to\infty} f(x) = \infty$ and
$$\lim_{x\to\infty} \frac{f(x)}{\ln(\ln(\ldots(\ln x)\ldots))} = 0$$
holds for all n (where n is the number of logarithm functions in the denominator)?

Asian Pacific Mathematical Olympiad 1989 (5)

Determine all functions f from the reals to the reals for which
    (i) $f(x)$ is strictly increasing,
    (ii) $f(x) + g(x) = 2x$ for all real x where $g(x)$ is the composition inverse function to $f(x)$. (Note: f and g are said to be composition inverses if $f(g(x)) = x$ and $g(f(x)) = x$ for all real x.)

IMO Proposal by Greece (1989) (6)

Let $g : \mathbf{C} \to \mathbf{C}$, $\omega \in \mathbf{C}$, $a \in \mathbf{C}$, with $\omega^3 = 1$ and $\omega \neq 1$. Show that there is one and only one function $f : \mathbf{C} \to \mathbf{C}$ such that
$$f(z) + f(\omega z + a) = g(z), \; z \in \mathbf{C}.$$
Find the function f.

11th Austrian-Polish Mathematics Competition (4)

Determine all strictly monotone increasing functions $f : \mathbf{R} \to \mathbf{R}$ satisfying the functional equation
$$f(f(x) + y) = f(x + y) + f(0)$$
for all $x, y \in \mathbf{R}$.

Australian Mathematical Olympiad 1985 (6)

Find all polynomials $f(x)$ with real coefficients such that
$$f(x) \cdot f(x+1) = f(x^2 + x + 1).$$

Singapore MSI Mathematical Competition 1988 (1)

Let $f(x)$ be a polynomial of degree n such that $f(k) = \frac{k}{k+1}$ for each $k = 0, 1, 2, \ldots, n$. Find $f(n+1)$.

First Selection Test of the Chinese I.M.O. Team 1988 (2)

Determine all functions f from the rational numbers to the complex numbers such that
    (i) $f(x_1 + x_2 + \cdots + x_{1988}) = f(x_1)f(x_2)\ldots f(x_{1988})$
for all rational numbers $x_1, x_2, \ldots, x_{1988}$, and
    (ii) $\overline{f(1988)}f(x) = f(1988)\overline{f(x)}$
for all rational numbers x, where $\overline{z}$ denotes the complex conjugate of z.

43rd Mathematical Olympiad (1991-92) in Poland (Final round) (2)

Determine all functions f defined on the set of positive rational numbers, taking values in the same set, which satisfy for every positive rational number x the conditions

$$f(x+1) = f(x) + 1 \quad \text{and} \quad f(x^3) = (f(x))^3.$$

43rd Mathematical Olympiad (1991-92) in Poland (Final round) (4)

Define the sequence of functions $f_0, f_1, f_2, \ldots$ by

$$f_0(x) = 8 \quad \text{for all } x \in \mathbb{R}.$$
$$f_{n+1}(x) = \sqrt{x^2 + 6f_n(x)} \quad \text{for } n = 0, 1, 2, \ldots \text{ and for all } x \in \mathbb{R}.$$

For every positive integer n, solve the equation $f_n(x) = 2x$.

Canadian Mathematical Olympiad 1996 (3)

We denote an arbitrary permutation of the integers $1, \ldots, n$ by $a_1, \ldots, a_n$. Let $f(n)$ be the number of these permutations such that
(i) $a_1 = 1$;
(ii) $|a_i - a_{i+1}| \leq 2, i = 1, \ldots, n-1$.
Determine whether $f(1996)$ is divisible by 3.

Canadian Mathematical Olympiad 1996 (5)

Let $r_1, r_2, \ldots, r_m$ be a given set of m positive rational numbers such that $\sum_{k=1}^{m} r_k = 1$. Define the function f by $f(n) = n - \sum_{k=1}^{m}[r_k n]$ for each positive integer n. Determine the minimum and maximum values of $f(n)$. Here $[x]$ denotes the greatest integer less than or equal to x.

10th Iranian Mathematical Olympiad (Second Stage Exam) (6)

Let X be a non-empty finite set and $f : X \to X$ a function such that for all x in X, $f^p(x) = x$, where p is a constant prime. If $Y = \{x \in X : f(x) \neq x\}$, prove that the number of elements of Y is divisible by p.

6th Korean Mathematical Olympiad (Final Round), 1993 (5)

Let n be a given natural number. Find all the continuous functions $f(x)$ satisfying:

$$\binom{n}{0}f(x) + \binom{n}{1}f(x^2) + \binom{n}{2}f(x^{2^2}) + \cdots + \binom{n}{n-1}f(x^{2^{n-1}}) + \binom{n}{n}f(x^{2^n})$$

6th Irish Mathematical Olympiad, 1993 (4)

Let $a_0, a_1, \ldots, a_{n-1}$ be real numbers, where $n \geq 1$, and let $f(x) = x^n + a_{n-1}x^{n-1} + \cdots + a_0$ be such that $|f(0)| = f(1)$ and each root α of f is real and satisfies $0 < \alpha < 1$. Prove that the product of the roots does not exceed $1/2^n$.

Turkish Mathematical Olympiad, 1993 (Final Selection Test) (6)

Let $\mathbb{Q}^+$ denote the set of all positive rational numbers. Find all functions $f : \mathbb{Q}^+ \to \mathbb{Q}^+$ such that

$$\text{for every } x, y \in \mathbb{Q}^+, \quad f\left(x + \frac{y}{x}\right) = f(x) + \frac{f(y)}{f(x)} + 2y.$$

Canadian Mathematical Olympiad, 1997 (5)

Write the sum

$$\sum_{k=0}^{n} \frac{(-1)^k \binom{n}{k}}{k^3 + 9k^2 + 26k + 24}.$$

in the form $\frac{p(n)}{q(n)}$, where p and q are polynomials with integer coefficients.

16th Austrian Polish Mathematical Competition (3)

Let the function f be defined as follows:

If $n = p^k > 1$ is a power of a prime number p, then $f(n) := n + 1$.

If $n = p_1^{k_1} \cdots p_r^{k_r}$ ($r > 1$) is a product of powers of pairwise different prime numbers, then $f(n) := p_1^{k_1} + \cdots + p_r^{k_r}$.

For every $m > 1$ we construct the sequence $\{a_0, a_1, \ldots\}$ such that $a_0 = m$ and $a_{j+1} = f(a_j)$ for $j \geq 0$. We denote by $g(m)$ the smallest element of this sequence. Determine the value of $g(m)$ for all $m > 1$.

Romanian First Team Selection Test, 1993 (34th IMO) (1)

Find the greatest real number a such that

$$\frac{x}{\sqrt{y^2 + z^2}} + \frac{y}{\sqrt{z^2 + x^2}} + \frac{z}{\sqrt{x^2 + y^2}} > a$$

is true for all positive real numbers x, y, z.

Romanian First Team Selection Test, 1993 (34th IMO) (4)

Show that for any function $f : \mathcal{P}(\{1, 2, \ldots, n\}) \to \{1, 2, \ldots, n\}$ there exist two subsets, A and B, of the set $\{1, 2, \ldots, n\}$, such that $A \neq B$ and $f(A) = f(B) = \max\{i \mid i \in A \cap B\}$.

Czechoslovakia Mathematical Olympiad, 1993 (5)

Find all functions $f : \mathbb{Z} \to \mathbb{Z}$ such that $f(-1) = f(1)$ and
$$f(x) + f(y) = f(x + 2xy) + f(y - 2xy)$$
for all integers x, y.

Mathematical Contest Baltic Way, 1992 (7)

Let $a = \sqrt[1992]{1992}$. Which number is greater:

$$\left.\begin{array}{r} a^{\cdot^{\cdot^{\cdot^{a}}}} \\ a^{a} \\ a \end{array}\right\} 1992 \quad \text{or} \quad 1992?$$

Mathematical Contest Baltic Way, 1992 (9)

A polynomial $f(x) = x^3 + ax^2 + bx + c$ is such that $b < 0$ and $ab = 9c$. Prove that the polynomial has three different real roots.

Mathematical Contest Baltic Way, 1992 (10)

Find all fourth degree polynomials $p(x)$ such that the following four conditions are satisfied:

(i) $p(x) = p(-x)$, for all x,

(ii) $p(x) \geq 0$, for all x,

(iii) $p(0) = 1$,

(iv) $p(x)$ has exactly two local minimum points x_1 and x_2 such that $|x_1 - x_2| = 2$.

Mathematical Contest Baltic Way, 1992 (11)

Let $\mathbb{Q}^+$ denote the set of positive rational numbers. Show that there exists one and only one function $f : \mathbb{Q}^+ \to \mathbb{Q}^+$ satisfying the following conditions:

(i) If $0 < q < \frac{1}{2}$ then $f(q) = 1 + f\left(\frac{q}{1-2q}\right)$.
(ii) If $1 < q \leq 2$ then $f(q) = 1 + f(q-1)$.
(iii) $f(q) \cdot f(\frac{1}{q}) = 1$ for all $q \in \mathbb{Q}^+$.

Mathematical Contest Baltic Way, 1992 (12)

Let $\mathbb{N}$ denote the set of positive integers. Let $\varphi : \mathbb{N} \to \mathbb{N}$ be a bijective function and assume that there exists a finite limit

$$\lim_{n \to \infty} \frac{\varphi(n)}{n} = L.$$

What are the possible values of L?

3rd Mathematical Olympiad of the Republic of China, 1994 (First day) (2)

Let a, b, c be positive real numbers, α be a real number. Suppose that
$$f(\alpha) = abc(a^\alpha + b^\alpha + c^\alpha)$$
$$g(\alpha) = a^{\alpha+2}(b+c-a) + b^{\alpha+2}(a-b+c) + c^{\alpha+2}(a+b-c)$$

Determine the magnitude between $f(\alpha)$ and $g(\alpha)$.

Israel Mathematical Olympiad, 1994 (1)

p and q are positive integers. f is a function defined for positive numbers and attains only positive values, such that $f(xf(y)) = x^p y^q$. Prove that $q = p^2$.

Israel Mathematical Olympiad, 1994 (5)

Find all real coefficients polynomials $p(x)$ satisfying

$$(x-1)^2 p(x) = (x-3)^2 p(x+2)$$

for all x.

Swedish Mathematical Olympiad, 1993 (6)

Let a and b be real numbers and let $f(x) = (ax+b)^{-1}$. For which a and b are there three distinct real numbers x_1, x_2, x_3 such that $f(x_1) = x_2$, $f(x_2) = x_3$ and $f(x_3) = x_1$?

Irish Mathematical Olympiad, 1994 (3)

Determine with proof all real polynomials $f(x)$ satisfying the equation
$$f(x^2) = f(x)f(x-1).$$

37th International Mathematical Olympiad, 1996 (Shortlist) (7)

Let f be a function from the set of real numbers $\mathbb{R}$ into itself such that for all $x \in \mathbb{R}$, we have $|f(x)| \leq 1$ and
$$f\left(x + \frac{13}{42}\right) + f(x) = f\left(x + \frac{1}{6}\right) + f\left(x + \frac{1}{7}\right).$$

Prove that f is a periodic function (that is, there exists a non-zero real number c such that $f(x+c) = f(x)$ for all $x \in \mathbb{R}$).

Croatian National Mathematics Competition, 1994 (4th Class) (2)

For a complex number z let $w = f(z) = \dfrac{2}{3-z}$.

(a) Determine the set $\{w : z = 2 + iy, y \in \mathbb{R}\}$ in the complex plane.

(b) Show that the function w can be written in the form
$$\frac{w-1}{w-2} = \lambda \frac{z-1}{z-2}.$$

(c) Let $z_0 = \frac{1}{2}$ and the sequence (z_n) be defined recursively by
$$z_n = \frac{2}{3 - z_{n-1}}, \quad n \geq 1.$$

Using the property (b) calculate the limit of the sequence (z_n).

Croatian National Mathematics Competition, 1994 (4th Class) (3)

Determine all polynomials $P(x)$ with real coefficients such that for some $n \in \mathbb{N}$ we have $xP(x-n) = (x-1)P(x)$, for all $x \in \mathbb{R}$.

17th Austrian-Polish Mathematics Competition, 1994 (1)

The function $f: \mathbb{R} \to \mathbb{R}$ satisfies for all $x \in \mathbb{R}$ the conditions
$$f(x+19) \leq f(x) + 19 \quad \text{and} \quad f(x+94) \geq f(x) + 94.$$

Show that $f(x+1) = f(x) + 1$ for all $x \in \mathbb{R}$.

37th International Mathematical Olympiad, 1996 (Shortlist) (6)

Let n be an even positive integer. Prove that there exists a positive integer k such that

$$k = f(x)(x+1)^n + g(x)(x^n + 1)$$

for some polynomials $f(x)$, $g(x)$ having integer coefficients. If k_0 denotes the least such k, determine k_0 as a function of n.

37th International Mathematical Olympiad, 1996 (Shortlist) (8)

Let the sequence $a(n)$, $n = 1, 2, 3, \ldots$, be generated as follows: $a(1) = 0$, and for $n > 1$,

$$a(n) = a(\lfloor n/2 \rfloor) + (-1)^{n(n+1)/2}.$$

(Here $\lfloor t \rfloor$ = the greatest integer $\leq t$.)

(a) Determine the maximum and minimum value of $a(n)$ over $n \leq 1996$, and find all $n \leq 1996$ for which these extreme values are attained.

(b) How many terms $a(n)$, $n \leq 1996$, are equal to 0?

Iranian National Mathematical Olympiad, 1994 (Second Round) (6)

$f(x)$ and $g(x)$ are polynomials with real coefficients such that for infinitely many rational values x, $\frac{f(x)}{g(x)}$ is rational. Prove that $\frac{f(x)}{g(x)}$ can be written as the ratio of two polynomials with rational coefficients.

Balkan Mathematical Olympiad, 1994 (2)

Show that the polynomial

$$x^4 - 1994x^3 + (1993 + m)x^2 - 11x + m, \quad m \in \mathbb{Z}$$

has at most one integral root.

38th Mathematics Competition of the Republic of Slovenia (4th Grade) (1)

Prove that there does not exist a function $f : \mathbb{Z} \to \mathbb{Z}$, for which $f(f(x)) = x + 1$ for every $x \in \mathbb{Z}$.

Israel Mathematical Olympiad, 1995 (6)

Let n be a given positive integer. A_n is the set of all points in the plane, whose x and y coordinates are positive integers between 0 and n. A point (i, j) is called "internal" if $0 < i, j < n$. A real function f, defined on A_n, is called a "good function" if it has the following property: for every internal point x, the value of $f(x)$ is the mean of its values on the four neighbouring points (the neighbouring points of x are the four points whose distance from x equals 1). f and g are two given good functions and $f(a) = g(a)$ for every point a in A_n which is not internal (that is, a boundary point). Prove that $f \equiv g$.

Israel Mathematical Olympiad, 1995 (10)

α is a given real number. Find all functions $f : (0, \infty) \mapsto (0, \infty)$ such that the equality

$$\alpha x^2 f\left(\frac{1}{x}\right) + f(x) = \frac{x}{x+1}$$

holds for all real $x > 0$.

45th Latvian Mathematical Olympiad, 1994 (12th Grade) (3)

Does there exist a polynomial $P(x, y)$ in two variables such that

(a) $P(x, y) > 0$ for all x, y?

(b) for each $c > 0$ there exist x and y such that $P(x, y) = c$?

4th Mathematical Olympiad of the Republic of China (Taiwan), 1995 (1)

Let $P(x) = a_0 + a_1 x + \cdots + a_{n-1} x^{n-1} + a_n x^n$ be a polynomial with complex coefficients. Suppose the roots of $P(x)$ are $\alpha_1, \alpha_2, \ldots, \alpha_n$ with $|\alpha_1| > 1$, $|\alpha_2| > 1$, $\ldots$, $|\alpha_j| > 1$, and $|\alpha_{j+1}| \leq 1$, $\ldots$, $|\alpha_n| \leq 1$. Prove:

$$\prod_{i=1}^{j} |\alpha_i| \leq \sqrt{|a_0|^2 + |a_1|^2 + \cdots + |a_n|^2}.$$

4th Mathematical Olympiad of the Republic of China (Taiwan), 1995 (4)

Given n distinct integers $m_1, m_2, \ldots, m_n$, prove that there exists a polynomial $f(x)$ of degree n and with integral coefficients which satisfies the following conditions:

(1) $f(m_i) = -1$, for all i, $1 \leq i \leq n$.

(2) $f(x)$ cannot be factorized into a product of two non-constant polynomials with integral coefficients.

18th Austrian-Polish Mathematics Competition, 1995 (3)

Let $P(x) = x^4 + x^3 + x^2 + x + 1$. Show that there exist polynomials $Q(y)$ and $R(y)$ of positive degrees, with integer coefficients, such that $Q(y) \cdot R(y) = P(5y^2)$ for all y.

31st Spanish Mathematical Olympiad, 1994 (First Round) (1)

Let a, b, c be distinct real numbers and $P(x)$ a polynomial with real coefficients. If:

- the remainder on division of $P(x)$ by $x - a$ equals a,
- the remainder on division of $P(x)$ by $x - b$ equals b,
- and the remainder on division of $P(x)$ by $x - c$ equals c;

determine the remainder on division of $P(x)$ by $(x - a)(x - b)(x - c)$.

31st Spanish Mathematical Olympiad, 1994 (First Round) (7)

Show that there exists a polynomial $P(x)$, with integer coefficients, such that $\sin 1°$ is a root of $P(x) = 0$.

Vietnamese Mathematical Olympiad, 1996 (Category A) (4)

Determine all functions $f : \mathbb{N}^* \to \mathbb{N}^*$ satisfying:

$$f(n) + f(n+1) = f(n+2)f(n+3) - 1996$$

for every $n \in \mathbb{N}^*$ ($\mathbb{N}^*$ is the set of positive integers).

Vietnamese Mathematical Olympiad, 1996 (Category B) (4)

Determine all functions $f : \mathbb{Z} \to \mathbb{Z}$ satisfying simultaneously two conditions:

(i) $f(1995) = 1996$

(ii) for every $n \in \mathbb{Z}$, if $f(n) = m$, then $f(m) = n$ and $f(m+3) = n - 3$, ($\mathbb{Z}$ is the set of integers).

19th Austrian-Polish Mathematics Competition, 1996 (3)

The polynomials $P_n(x)$ are defined recursively by $P_0(x) = 0$, $P_1(x) = x$ and

$$P_n(x) = xP_{n-1}(x) + (1-x)P_{n-2}(x) \quad \text{for} \quad n \geq 2.$$

For every natural number $n \geq 1$, find all real numbers x satisfying the equation $P_n(x) = 0$.

3rd Turkish Mathematical Olympiad, 1995 (3)

Let $\mathbb{N}$ denote the set of positive integers. Let A be a real number and $(a_n)_{n=1}^{\infty}$ be a sequence of real numbers such that $a_1 = 1$ and

$$1 < \frac{a_{n+1}}{a_n} \leq A \quad \text{for all} \quad n \in \mathbb{N}.$$

(a) Show that there is a unique non-decreasing surjective function $k : \mathbb{N} \to \mathbb{N}$ such that $1 < \frac{A^{k(n)}}{a_n} \leq A$ for all $n \in \mathbb{N}$.

(b) If k takes every value at most m times, show that there exists a real number $C > 1$ such that $C^n \leq A a_n$ for all $n \in \mathbb{N}$.

3rd Turkish Mathematical Olympiad, 1995 (6)

Let $\mathbb{N}$ denote the set of positive integers. Find all surjective functions $f : \mathbb{N} \to \mathbb{N}$ satisfying the condition

$$m \mid n \iff f(m) \mid f(n)$$

for all $m, n \in \mathbb{N}$.

Australian Mathematical Olympiad, 1996 (2)

Let $p(x)$ be a cubic polynomial with roots r_1, r_2, r_3. Suppose that

$$\frac{p\left(\frac{1}{2}\right) + p\left(-\frac{1}{2}\right)}{p(0)} = 1000.$$

Find the value of $\frac{1}{r_1 r_2} + \frac{1}{r_2 r_3} + \frac{1}{r_3 r_1}$.

Australian Mathematical Olympiad, 1996 (8)

Let f be a function that is defined for all integers and takes only the values 0 and 1. Suppose f has the following properties:

(i) $f(n + 1996) = f(n)$ for all integers n;

(ii) $f(1) + f(2) + \cdots + f(1996) = 45$.

Prove that there exists an integer t such that $f(n+t) = 0$ for all n for which $f(n) = 1$ holds.

47th Polish Mathematical Olympiad, 1995 (1)

Find all pairs (n, r), with n a positive integer, r a real number, for which the polynomial $(x+1)^n - r$ is divisible by $2x^2 + 2x + 1$.

10th Nordic Mathematical Contest, 1996 (4)

A real-valued function f is defined for positive integers, and a positive integer a satisfies

$$f(a) = f(1995), \quad f(a+1) = f(1996), \quad f(a+2) = f(1997),$$

$$f(n+a) = \frac{f(n) - 1}{f(n) + 1} \quad \text{for any positive integer } n.$$

(a) Prove that $f(n + 4a) = f(n)$ for any positive integer n.

(b) Determine the smallest possible value of a.

9th Irish Mathematical Olympiad, 1996 (1)

For each positive integer n, let $f(n)$ denote the greatest common divisor of $n! + 1$ and $(n+1)!$ (where ! denotes "factorial"). Find, with proof, a formula for $f(n)$ for each n.

9th Irish Mathematical Olympiad, 1996 (3)

Let K be the set of all real numbers x with $0 \leq x \leq 1$. Let f be a function from K to the set of all real numbers $\mathbb{R}$ with the following properties:

(i) $f(1) = 1$.

(ii) $f(x) \geq 0$ for all $x \in K$.

(iii) if x, y and $x + y$ are all in K, then

$$f(x + y) \geq f(x) + f(y).$$

Prove that $f(x) \leq 2x$ for all $x \in K$.

St. Petersburg City Mathematical Olympiad, 1996 (Third Round) (5)

Find all quadruplets of polynomials $p_1(x), p_2(x), p_3(x), p_4(x)$ with real coefficients possessing the following remarkable property: for all integers x, y, z, t satisfying the condition $xy - zt = 1$, the equality $p_1(x)p_2(y) - p_3(z)p_4(t) = 1$ holds.

Republic of Moldova XL Mathematical Olympiad, 1996 (11-12) (5)

Let p be the number of functions defined on the set $\{1, 2, \ldots, m\}$, $m \in N^*$, with values in the set $\{1, 2, \ldots, 35, 36\}$ and q be the number of functions defined on the set $\{1, 2, \ldots, n\}$, $n \in N^*$, with values in the set $\{1, 2, 3, 4, 5\}$. Find the least possible value for the expression $|p - q|$.

Ukrainian Mathematical Olympiad, 1996 (7)

Does a function $f : \mathbb{R} \to \mathbb{R}$ exist which is not a polynomial and such that for all real x

$$(x-1)f(x+1) - (x+1)f(x-1) = 4x(x^2-1)?$$

Taiwan Mathematical Olympiad, 1996 (6)

Let $q_0, q_1, q_2, \ldots$ be a sequence of integers such that
(a) for any $m > n$, $m - n$ is a factor of $q_m - q_n$, and
(b) $|q_n| \leq n^{10}$ for all integers $n \geq 0$.

Show that there exists a polynomial $Q(x)$ satisfying $Q(n) = q_n$ for all n.

Croatian Mathematical Olympiad, 1995 (IV Class) (3)

Determine all functions $f : \mathbb{R} \to \mathbb{R}$ continuous at 0, which satisfy the following relation

$$f(x) - 2f(tx) + f(t^2x) = x^2 \quad \text{for all} \quad x \in \mathbb{R},$$

where $t \in (0,1)$ is a given number.

13th Iranian Mathematical Olympiad 1995 (3)

Let $P(x)$ be a polynomial with rational coefficients such that $P^{-1}(\mathbb{Q}) \subseteq \mathbb{Q}$. Show that P is linear.

13th Iranian Mathematical Olympiad 1995 (5)

Does there exist a function $f : \mathbb{R} \to \mathbb{R}$ that fulfils all of the following conditions:

(a) $f(1) = 1$
(b) there exists $M > 0$ such that $-M < f(x) < M$
(c) if $x \neq 0$ then

$$f\left(x + \frac{1}{x^2}\right) = f(x) + \left(f\left(\frac{1}{x}\right)\right)^2 ?$$

Estonian Mathematical Contest, 1995 (Final Round) (3)

Prove that the polynomial $P_n(x) = 1 + x + \frac{x^2}{2} + \frac{x^3}{6} + \cdots + \frac{x^n}{n!}$ has no zeros if n is even and has exactly one zero if n is odd.

Estonian Mathematical Contest, 1995 (Final Round) (5)

Find all functions $f : \mathbb{R} \to \mathbb{R}$ satisfying the following conditions for all $x \in \mathbb{R}$.

(a) $f(x) = -f(-x)$;
(b) $f(x+1) = f(x) + 1$;
(c) $f\left(\dfrac{1}{x}\right) = \dfrac{1}{x^2} f(x)$, if $x \neq 0$.

6th ROC Taiwan Mathematical Olympiad, 1997 (Part I) (1)

Let a be a rational number, b, c, d be real, and the function $f : R \to [-1,1]$ satisfying

$$f(x+a+b) - f(x+b) = c \cdot \lfloor x + 2a + \lfloor x \rfloor - 2\lfloor x + a \rfloor - \lfloor b \rfloor \rfloor + d$$

for each $x \in R$, where $\lfloor t \rfloor$ denotes the largest integer that is less than or equal to t. Show that f is a periodic function (that is, there is a positive number p such that $f(x+p) = f(x)\ \forall x \in R$).

6th ROC Taiwan Mathematical Olympiad, 1997 (Part III) (1)

Determine all the possible integers k such that there is a function $f : \mathbb{N} \longrightarrow \mathbb{Z}$ such that

(i) $f(1997) = 1998$,
(ii) $f(ab) = f(a) + f(b) + k \cdot f(d(a,b))$, $\forall a, b \in \mathbb{N}$, where $d(a,b)$ denotes the greatest common divisor of a and b.

XXXIII Spanish Mathematical Olympiad, 1996 (8)

For each real number x, we denote by $\lfloor x \rfloor$ the biggest integer which is less than or equal to x. We define

$$q(n) = \left\lfloor \frac{n}{\lfloor \sqrt{n} \rfloor} \right\rfloor, \quad n = 1, 2, 3, \ldots.$$

(a) Forming a table with the values of $q(n)$ for $1 \leq n \leq 25$, make a conjecture about the numbers n for which $q(n) > q(n+1)$.

(b) Determine, with reasons, all the positive integer n such that

$$q(n) > q(n+1).$$

20th Austrian-Polish Mathematical Competition, 1997 (5)

Let p_1, p_2, p_3, and p_4 be four distinct prime numbers. Prove that there does not exist a cubic polynomial $Q(x) = ax^3 + bx^2 + cx + d$ with integer coefficients such that

$$|Q(p_1)| = |Q(p_2)| = |Q(p_3)| = |Q(p_4)| = 3.$$

Chinese Mathematical Olympiad, 1997 (1)

Let $x_1, x_2, \ldots, x_{1997}$ be real numbers satisfying the following two conditions:

(a) $-\frac{1}{\sqrt{3}} \leq x_i \leq \sqrt{3}$ $(i = 1, 2, \ldots, 1997)$;

(b) $x_1 + x_2 + \cdots + x_{1997} = -318\sqrt{3}$.

Find the maximum of $x_1^{12} + x_2^{12} + \cdots + x_{1997}^{12}$ and give your reason.

Swedish Mathematical Competition, 1996 (Final Round) (3)

For all integers $n \geq 1$ the functions p_n are defined for $x \geq 1$ by

$$p_n(x) = \frac{1}{2}\left(\left(x + \sqrt{x^2 - 1}\right)^n + \left(x - \sqrt{x^2 - 1}\right)^n\right).$$

Show that $p_n(x) \geq 1$ and that $p_{mn}(x) = p_m(p_n(x))$.

Latvian Mathematical Olympiad, 1997 (1st TST) (2)

Does there exist a function $f(x)$ which is defined for all reals and for which the identities

$$f(f(x)) = x \quad \text{and} \quad f(f(x) + 1) = 1 - x$$

hold?

Mathematical Olympiad in Bosnia and Herzegovina, 1997 (1st Day) (3)

Let $f : A \to \mathbb{R}$, $A \subseteq \mathbb{R}$, be a function with the following characteristic:
$$f(x+y) = f(x) \cdot f(y) - f(xy) + 1, \quad (\forall x, y \in A).$$

(a) If $f : A \to \mathbb{R}$, $\mathbb{N} \subset A \subseteq \mathbb{R}$, is such a function, prove that the following is true:
$$f(n) = \begin{cases} \frac{c^{n+1}-1}{c-1}, & \forall n \in \mathbb{N},\ c \neq 1, \\ n+1, & \forall n \in \mathbb{N},\ c = 1, \end{cases}$$
$(c = f(1) - 1)$.

(b) Solve the given functional equation for $A = \mathbb{N}$.

(c) If $A = \mathbb{Q}$, find all the functions f which satisfy the given equation and the condition $f(1997) \neq f(1998)$.

Fourth National Mathematical Olympiad of Turkey, 1997 (5)

Let $\mathbb{R}$ stand for the set of all real numbers. Show that there is no function $f : \mathbb{R} \to \mathbb{R}$ such that
$$f(x+y) > f(x)(1+yf(x))$$
for all positive real x, y.

20th Austrian-Polish Mathematical Competition, 1997 (6)

Prove that there does not exist a function $f : \mathbb{Z} \to \mathbb{Z}$ such that $f(x + f(y)) = f(x) - y$ for all integers x and y.

Estonian Mathematical Olympiad, 1997 (Final Round) (7)

A function f satisfies the condition
$$f(1) + f(2) + \cdots + f(n) = n^2 f(n)$$
for any positive integer n. Given that $f(1) = 999$, find $f(1997)$.

Ukrainian Mathematical Olympiad, 1997 (3)

Let $d(n)$ denote the greatest odd divisor of the natural number n. We define the function $f : \mathbb{N} \to \mathbb{N}$ as follows: $f(2n-1) = 2^n$, $f(2n) = n + \frac{2n}{d(n)}$ for all $n \in \mathbb{N}$.
Find all k such that $f(f(\ldots f(1)\ldots)) = 1997$, where f is iterated k times.

Ukrainian Mathematical Olympiad, 1997 (6)

Let $\mathbb{Q}^+$ denote the set of all positive rational numbers.
Find all functions $f : \mathbb{Q}^+ \to \mathbb{Q}^+$ such that for all $x \in \mathbb{Q}^+$:

(a) $f(x+1) = f(x)+1$,

(b) $f(x^2) = (f(x))^2$.

36th Armenian National Mathematical Olympiad, 1997 (1)

Let
$$p(x) = (x-a_1)^{n_1}(x-a_2)^{n_2}(x-a_3)^{n_3}$$
be a polynomial, such that
$$p(x) - 1 = (x-b_1)^{k_1}(x-b_2)^{k_2}(x-b_3)^{k_3},$$
where the numbers a_1, a_2, a_3, as well as b_1, b_2, b_3, are distinct, and $n_1, n_2, n_3, k_1, k_2, k_3$ are natural numbers. Prove that the degree of the polynomial $p(x)$ does not exceed 5.

38th IMO Croation Team Selection Test, 1997 (2)

Let a, b, c, d be real numbers such that at least one is different from zero. Prove that all roots of the polynomial
$$P(x) = x^6 + ax^3 + bx^2 + cx + d$$
cannot be real.

Vietnamese Team Selection Test, 1997 (4)

Let $f : \mathbb{N} \to \mathbb{Z}$ be the function defined by:
$$f(0) = 2, \quad f(1) = 503,$$
$$f(n+2) = 503 f(n+1) - 1996 f(n) \quad \text{for all } n \in \mathbb{N}.$$

For every $k \in \mathbb{N}^*$, take k arbitrary integers $s_1, s_2, \ldots, s_k$ such that $s_i \geq k$ for all $i = 1, 2, \ldots, k$, and for every s_i ($i = 1, 2, \ldots, k$), take an arbitrary prime divisor $p(s_i)$ of $f(2^{s_i})$.

Prove that for positive integers $t \leq k$, we have:
$$\sum_{i=1}^{k} p(s_i) \mid 2^t \quad \text{if and only if} \quad k \mid 2^t.$$

Vietnamese Mathematical Olympiad, 1997 (3)

How many functions $f : \mathbb{N}^* \to \mathbb{N}^*$ are there that simultaneously satisfy the two following conditions:

(i) $f(1) = 1$,

(ii) $f(n) \cdot f(n+2) = (f(n+1))^2 + 1997$ for all $n \in \mathbb{N}^*$?

($\mathbb{N}^*$ denotes the set of all positive integers.)

Vietnamese Mathematical Olympiad, 1997 (4)

(a) Find all polynomials of least degree, with rational coefficients, such that
$$f(\sqrt[3]{3} + \sqrt[3]{9}) = 3 + \sqrt[3]{3}.$$

(b) Does there exist a polynomial with integer coefficients such that
$$f(\sqrt[3]{3} + \sqrt[3]{9}) = 3 + \sqrt[3]{3}?$$

28th Austrian Mathematical Olympiad, 1997 (6)

Let n be a fixed natural number. Determine all polynomials $x^2 + ax + b$, where $a^2 \geq 4b$, such that $x^2 + ax + b$ divides $x^{2n} + ax^n + b$.

Iranian Mathematical Olympiad, 1997 (Second Round) (4)

Find all functions $f : \mathbb{N} \to \mathbb{N} \setminus \{1\}$ such that for all $n \in \mathbb{N} \setminus \{0\}$ we have,
$$f(n+1) + f(n+3) = f(n+5)f(n+7) - 1375.$$

Iranian Mathematical Olympiad, 1997 (Final Round) (5)

Suppose that $f : \mathbb{R}^+ \to \mathbb{R}^+$ is a decreasing continuous function that fulfills the following condition for all $x, y \in \mathbb{R}^+$:
$$f(x+y) + f(f(x) + f(y)) = f\big(f(x + f(y)) + f(y + f(x))\big).$$

Prove that $f(x) = f^{-1}(x)$.

Swedish Mathematical Competition, 1997 (Final Round) (3)

Let the sum of the two integers A and B be odd. Show that any integer can be written in the form $x^2 - y^2 + Ax + By$, where x and y are integers.

Swedish Mathematical Competition, 1997 (Final Round) (5)

Let $s(m)$ denote the sum of the digits of the integer m. Prove that for any integer n, with $n > 1$ and $n \neq 10$, there is a unique integer $f(n) \geq 2$ such that $s(k) + s(f(n) - k) = n$ for all integers k satisfying $0 < k < f(n)$.

Ukrainian Mathematical Olympiad, 1998 (11th Grade) (6)

The function $f(x)$ is defined on $[0, 1]$ and has values in $[0, 1]$. It is known that $\lambda \in (0, 1)$ exists such that $f(\lambda) \neq 0$ and $f(\lambda) \neq \lambda$. Also
$$f(f(x) + y) = f(x) + f(y)$$
for all x and y from the range of definition of the equality.

(a) Give an example of such a function.

(b) Prove that for any $x \in [0, 1]$,
$$\underbrace{f(f(\ldots f(x)\ldots))}_{19} = \underbrace{f(f(\ldots f(x)\ldots))}_{98}.$$

Vietnamese Mathematical Olympiad, 1998 (Category A, Day 1) (1)

Let $a \geq 1$ be a real number. Define a sequence $\{x_n\}$ ($n = 1, 2, \ldots$) of real numbers by
$$x_1 = a, \quad x_{n+1} = 1 + \ln\left(\frac{x_n^2}{1 + \ln x_n}\right).$$
Prove that the sequence $\{x_n\}$ has a finite limit, and determine it.

Vietnamese Mathematical Olympiad, 1998 (Category B, Day 1) (6)

Prove that for each positive odd integer n there is exactly one polynomial $P(x)$ of degree n with real coefficients satisfying
$$P\left(x - \frac{1}{x}\right) = x^n - \frac{1}{x^n}$$
for all real $x \neq 0$.
Determine if the above assertion holds for positive even integers n.

1st Mediterranean Mathematical Olympiad, 1998 (2)

(a) Prove that the polynomial $z^{2n} + z^n + 1$, $n \in \mathbb{N}$, is divisible by the polynomial $z^2 + z + 1$ if and only if n is not a multiple of 3.

(b) Find the necessary and sufficient condition that the natural numbers p, q must satisfy for the polynomial $z^p + z^q + 1$ to be divisible by $z^2 + z + 1$.

Final National Selection Competition for Greek Team, 1998 (4)

(a) A polynomial $P(x)$ with integer coefficients takes the value -2 for seven distinct integer values of x. Prove that it cannot take the value 1996.

(b) Prove that there are irrational numbers x, y such that the number x^y is rational.

Hungary-Israel Mathematical Competition, 1999 (1)

Let $f(x)$ be a polynomial whose degree is at least 2. Define the sequence $g_i(x)$ by: $g_1(x) = f(x)$ and $g_{n+1}(x) = f(g_n(x))$ for $n = 1, 2, \ldots$. Let r_n be the average of the roots of $g_n(x)$. It is given that $r_{19} = 99$. Find r_{99}.

Hungary-Israel Mathematical Competition, 1999 (3)

Find all the functions f from the set of rational numbers to the set of real numbers such that for all rational x, y,

$$f(x+y) = f(x)f(y) - f(xy) + 1.$$

Hungary-Israel Mathematical Competition, 1999 (5)

The function

$$f(x, y, z) = \frac{x^2 + y^2 + z^2}{x + y + z}$$

is defined for every x, y, z such that $x + y + z \neq 0$. Find a point (x_0, y_0, z_0) such that $0 < x_0^2 + y_0^2 + z_0^2 < \frac{1}{1999}$ and $1.999 < f(x_0, y_0, z_0) < 2$.

12th Korean Mathematical Olympiad, 1999 (2)

Suppose $f(x)$ is a function satisfying $|f(m+n) - f(m)| \leq \frac{n}{m}$ for all rational numbers n and m. Show that for all natural numbers k

$$\sum_{i=1}^{k} |f(2^k) - f(2^i)| \leq \frac{k(k-1)}{2}.$$

12th Korean Mathematical Olympiad, 1999 (4)

Suppose that for any real x ($|x| \neq 1$), a function $f(x)$ satisfies

$$f\left(\frac{x-3}{x+1}\right) + f\left(\frac{3+x}{1-x}\right) = x.$$

Find all possible $f(x)$.

Grosman Memorial Mathematical Olympiad, 1999 (4)

Consider a polynomial $f(x) = x^4 + ax^3 + bx^2 + cx + d$ with integer coefficients a, b, c, d. Prove that if $f(x)$ has exactly one real root then $f(x)$ can be factored into terms with rational coefficients.

Russian Mathematical Olympiad, 1999 (11th Form) (2)

A function $f : \mathbb{Q} \to \mathbb{Z}$ is considered. Prove that there exist two rational numbers a and b such that
$$\frac{f(a) + f(b)}{2} \leq f\left(\frac{a+b}{2}\right).$$

16th Iranian Mathematical Olympiad, 1999 (Second Round) (4)

Find all functions $f : \mathbb{R} \to \mathbb{R}$ satisfying,
$$f(f(x) + y) = f(x^2 - y) + 4f(x)y,$$
for all real numbers $x, y \in \mathbb{R}$.

Vietnamese Mathematical Olympiad, 1999 (Category B) (4)

Let $f(x)$ be a continuous function defined on $[0, 1]$ such that
(i) $f(0) = f(1) = 0$,
(ii) $2f(x) + f(y) = 3f\left(\frac{2x+y}{3}\right)$ $\forall\, x, y \in [0, 1]$.
Prove that $f(x) = 0$ for all $x \in [0, 1]$.

Turkish Team Selection Test for 40th IMO, 1999 (3)

Determine all functions $f : \mathbb{R} \longrightarrow \mathbb{R}$ such that the set
$$\left\{\frac{f(x)}{x} : x \neq 0 \quad \text{and} \quad x \in \mathbb{R}\right\}$$
is finite, and for all $x \in \mathbb{R}$
$$f(x - 1 - f(x)) = f(x) - x - 1.$$

Japanese Mathematical Olympiad, 1999 (Final Round) (2)

Let $f(x) = x^3 + 17$. Prove that for each natural number n, $n \geq 2$, there is a natural number x, for which $f(x)$ is divisible by 3^n but not by 3^{n+1}.

Japanese Mathematical Olympiad, 1999 (Final Round) (4)

Prove that
$$f(x) = (x^2 + 1^2)(x^2 + 2^2)(x^2 + 3^2) \cdots (x^2 + n^2) + 1$$
cannot be expressed as a product of two integral-coefficient polynomials with degree greater than 1.

Swiss Mathematical Contest, 1999 (First Day) (3)

Determine all functions $f : \mathbb{R} \setminus \{0\} \to \mathbb{R}$, satisfying
$$\frac{1}{x} f(-x) + f\left(\frac{1}{x}\right) = x \text{ for all } x \in \mathbb{R} \setminus \{0\}.$$

Swiss Mathematical Contest, 1999 (Second Day) (4)

Prove that for every polynomial $P(x)$ of degree 10 with integer coefficients there is an (in both directions) infinite arithmetic progression which does not contain $P(k)$ for any integer k.

St. Petersburg Mathematical Contest (25)

The sum of two continuous periodic functions is a non-constant continuous periodic function. Prove that the periods of these two functions are integral multiples of the period of their sum.

St. Petersburg Mathematical Contest (45)

Let $P(z)$ and $Q(z)$ be complex polynomials, one of which is not constant. Every root of $P(z)$ is also a root of $Q(z)$ and vice versa. Every root of $P(z) - 1$ is also a root of $Q(z) - 1$ and vice versa. Prove that $P = Q$.

Ukranian Mathematical Olympiad, 1999 (10th Grade) (5)

Let $P(x)$ be a polynomial with integer coefficients.
The sequence of integers $x_1, x_2, \ldots, x_n, \ldots$ satisfies the conditions $x_1 = x_{2000} = 1999$, $x_{n+1} = P(x_n)$, $n \geq 1$. Find the value of
$$\frac{x_1}{x_2} + \frac{x_2}{x_3} + \cdots + \frac{x_{1999}}{x_{2000}}.$$

XLIII Mathematical Olympiad of Moldova, 1999 (10th Form) (1)

Let the function $f : \mathbb{R} \to \mathbb{R}$, $f(x) = x^2 - 2ax - a^2 - \frac{3}{4}$, be considered. Find the values a for which the inequality $|f(x)| \leq 1$ is true for every $x \in [0, 1]$.

XLIII Mathematical Olympiad of Moldova, 1999 (10th Form) (5)

Find all the functions $f : \mathbb{R} \to \mathbb{R}$, which satisfy the relation
$$x \cdot f(x) = \lfloor x \rfloor \cdot f(\{x\}) + \{x\} \cdot f(\lfloor x \rfloor), \quad \forall\, x \in \mathbb{R},$$
where $\lfloor \cdot \rfloor$ and $\{\cdot\}$ denote the integral part and fractional part functions, respectively.

XLIII Mathematical Olympiad of Moldova, 1999 (10th Form) (6)

Find a polynomial of degree 3 with real coefficients such that each of its roots is equal to the square of one root of the polynomial $P(X) = X^3 + 9X^2 + 9X + 9$.

XLIII Mathematical Olympiad of Moldova, 1999 (11th Form) (2)

Let the number $n \in \mathbb{N}^*$ be given. Denote by M the set of all real numbers x for which there exists a finite sequence (a_p), $p = 1, \ldots, n$, with $a_p \in \{0, 1\}$, $p = 1, \ldots, n$, such that
$$x = 2^{-1} \cdot a_1 + 2^{-2} \cdot a_2 + \cdots + 2^{-n} \cdot a_n.$$

(a) Determine the set M, and prove that for every number $x \in M$ there exists a unique finite sequence (a_p), $p = 1, \ldots, n$, with the mentioned property.

(b) Find the function $f : M \to \mathbb{R}$ such that if (a_p) is the sequence defined above by the number x, then
$$f(x) = 2^{-1} \cdot 2000^{a_1} + 2^{-2} \cdot 2000^{a_2} + \cdots + 2^{-n} \cdot 2000^{a_n}, \quad \forall x \in M.$$

Italian Team Selection Test, 1999 (3)

(a) Determine all the strictly monotone functions $f : \mathbb{R} \to \mathbb{R}$ such that
$$f(x + (f(y)) = f(x) + y, \quad \forall\, x, y \in \mathbb{R}.$$

(b) Prove that for every integer $n > 1$ there do not exist strictly monotone functions $f : \mathbb{R} \to \mathbb{R}$ such that
$$f(x + f(y)) = f(x) + y^n, \quad \forall\, x, y \in \mathbb{R}.$$

Estonian Mathematical Contest, 1996 (3)

Prove that the polynomial $P_n(x) = 1 + x + \frac{x^2}{2} + \frac{x^3}{6} + \cdots + \frac{x^n}{n!}$ has no zeroes if n is even and has exactly one zero if n is odd.

Mongolian Team Selection Test for 40th IMO, 1999 (1)

Let n be a positive integer and $P(x)$ a polynomial of degree $2n$ such that $P(0) = 1$ and $P(k) = 2^{k-1}$ for $k = 1, 2, \ldots, 2n$. Prove that $2P(2n+1) - P(2n+2) = 1$.

Korean Mathematical Olympiad, 2000 (2)

Determine all functions f from the set of real numbers to itself such that for every x and y,
$$f(x^2 - y^2) = (x - y)(f(x) + f(y)).$$

Vietnamese Mathematical Olympiad, 2000 (4)

For every integer $n \geq 3$ and any given angle α in $(0, \pi)$, let $P_n(x) = x^n \sin \alpha - x \sin n\alpha + \sin(n-1)\alpha$.

(a) Prove that there is only one polynomial of the form $f(x) = x^2 + ax + b$ such that for every $n \geq 3$, $P_n(x)$ is divisible by $f(x)$.

(b) Prove that there does not exist a polynomial $g(x)$ of the form $g(x) = x + c$ such that for every $n \geq 3$, $P_n(x)$ is divisible by $g(x)$.

Bulgarian Mathematical Olympiad, 2000 (4)

Find all polynomials $P(x)$ with real coefficients such that we have $P(x)P(x+1) = P(x^2)$ for all real x.

Bulgarian Mathematical Olympiad, 2000 (6)

Let $\mathcal{A}$ be the set of all binary sequences of length n, and let $0 \in \mathcal{A}$ be the sequence all terms of which are zeroes. The sequence $c = \langle c_1, c_2, \ldots, c_n \rangle$ is called the sum of $a = \langle a_1, a_2, \ldots, a_n \rangle$ and $b = \langle b_1, b_2, \ldots, b_n \rangle$ if $c_i = 0$ when $a_i = b_i$ and $c_i = 1$ when $a_i \neq b_i$. Let $f : \mathcal{A} \to \mathcal{A}$ be a function such that $f(0) = 0$ and if the sequences a and b differ in exactly k terms then the sequences $f(a)$ and $f(b)$ differ also exactly in k terms. Prove that if a, b, and c are sequences from $\mathcal{A}$ such that $a + b + c = 0$, then $f(a) + f(b) + f(c) = 0$.

Taiwanese Mathematical Olympiad, 2000 (6)

Let f be a function from the set of positive integers to the set of non-negative integers such that $f(1) = 0$ and
$$f(n) = \max\{f(j) + f(n-j) + j\}$$
for all $n \geq 2$. Determine $f(2000)$.

Iranian Mathematical Olympiad, 2000 (6)

Prove that for every positive integer n, there exists a polynomial $p(x)$ with integer coefficients such that $p(1), p(2), \ldots, p(n)$ are distinct powers of 2.

Shortlist for IMO, 2000 (Belarus) (12)

Find all pairs of functions f and g from the set of real numbers to itself such that $f(x + g(y)) = xf(y) - yf(x) + g(x)$ for all real numbers x and y.

Shortlist for IMO, 2000 (United Kingdom) (20)

A function F is defined from the set of non-negative integers to itself such that, for every non-negative integer n, $F(4n) = F(2n) + F(n)$, $F(4n + 2) = F(4n) + 1$, and $F(2n + 1) = F(2n) + 1$. Prove that, for each positive integer m, the number of integers n with $0 \leq n < 2^m$ and $F(4n) = F(3n)$ is $F(2^{m+1})$.

32nd Austrian Mathematical Olympiad (4)

Determine all functions $f : \mathbb{R} \mapsto \mathbb{R}$, such that for all real numbers x and y the functional equation $f(f(x)^2 + f(y)) = x \cdot f(x) + y$ is satisfied.

14th Nordic Mathematical Contest (4)

The real-valued function f is defined for $0 \leq x \leq 1$, and satisfies $f(0) = 0$, $f(1) = 1$, and
$$\frac{1}{2} \leq \frac{f(z) - f(y)}{f(y) - f(x)} \leq 2,$$
for all $0 \leq x < y < z \leq 1$ with $z - y = y - x$. Prove that
$$\frac{1}{7} \leq f\left(\frac{1}{3}\right) \leq \frac{4}{7}.$$

Ukrainian Mathematical Olympiad, 2001 (Grade 11) (5)

Does there exist a function $f : \mathbb{R} \to \mathbb{R}$ such that for all $x, y \in \mathbb{R}$ the following equality holds?
$$f(xy) = \max\{f(x), y\} + \min\{f(y), x\}.$$

Hungary-Israel Mathematical Competition, 2001 (Individual) (3)

Find all continuous functions $f : \mathbb{R} \to \mathbb{R}$ such that, for all real x,
$$f(f(x)) = f(x) + x.$$

Hungary-Israel Mathematical Competition, 2001 (Individual) (4)

Let $P(x) = x^3 - 3x + 1$. Find the polynomial Q whose roots are the fifth power of the roots of P.

2nd Hong Kong Mathematical Olympiad, 1999 (4)

Determine all functions $f : \mathbb{R} \to \mathbb{R}$ such that, for all $x, y \in \mathbb{R}$,
$$f(x + yf(x)) = f(x) + xf(y).$$

17th Balkan Mathematical Olympiad, 2000 (1)

Find all the functions $f : \mathbb{R} \to \mathbb{R}$ with the property that
$$f(xf(x) + f(y)) = (f(x))^2 + y,$$
for any real numbers x and y.

49th Mathematical Olympiad of Lithuania, 2000 (6)

A function $f : \mathbb{R} \to \mathbb{R}$ satisfies the following equation for all real x and y:
$$(x + y)(f(x) - f(y)) = f(x^2) - f(y^2).$$
Find: (a) one such function; (b) all such functions.

XXXVI Spanish Mathematical Olympiad, 2000 (1)

Let $P(x) = x^4 + ax^3 + bx^2 + cx + 1$ and $Q(x) = x^4 + cx^3 + bx^2 + ax + 1$, with a, b, c real numbers and $a \neq c$. Find conditions on $a, b,$ and c so that $P(x)$ and $Q(x)$ have two common roots. In this case, solve the equations $P(x) = 0$, $Q(x) = 0$.

XXXVI Spanish Mathematical Olympiad, 2000 (6)

Show that there is no function $f : \mathbb{N} \to \mathbb{N}$ such that $f(f(n)) = n + 1$.

8th Macedonian Mathematical Olympiad (2)

Does there exist a function $f : \mathbb{N} \to \mathbb{N}$ such that for every $n \geq 2$,
$$f(f(n-1)) = f(n+1) - f(n)?$$

13th Irish Mathematical Olympiad (3)

Let $f(x) = 5x^{13} + 13x^5 + 9ax$. Find the least positive integer a such that 65 divides $f(x)$ for every integer x.

13th Irish Mathematical Olympiad (10)

Let $p(x) = a_0 + a_1 x + \cdots + a_n x^n$ be a polynomial with non-negative real coefficients. Suppose that $p(4) = 2$ and $p(16) = 8$. Prove that $p(8) \leq 4$, and find, with proof, all such polynomials with $p(8) = 4$.

Singapore Mathematical Olympiad, 2002 (Open Section, Part A) (1)

Let $f(x)$ be a function which satisfies
$$f(29 + x) = f(29 - x),$$
for all values of x. If $f(x)$ has exactly three real roots α, β, and γ, determine the value of $\alpha + \beta + \gamma$.

Singapore Mathematical Olympiad, 2002 (Open Section, Part A) (5)

It is given that the polynomial $p(x) = x^3 + ax^2 + bx + c$ has three distinct positive integer roots and $p(2002) = 2001$. Let $q(x) = x^2 - 2x + 2002$. It is also given that the polynomial $p(q(x))$ has no real roots. Determine the value of a.

Singapore Mathematical Olympiad, 2002 (Open Section, Part B) (4)

Find all real-valued functions $f : \mathbb{Q} \to \mathbb{R}$ defined on the set of all rational numbers $\mathbb{Q}$ satisfying the conditions
$$f(x + y) = f(x) + f(y) + 2xy,$$
for all x, y in $\mathbb{Q}$ and $f(1) = 2002$. Justify your answers.

15th Korean Mathematical Olympiad (2)

Find all functions $f : \mathbb{R} \to \mathbb{R}$ satisfying $f(x - y) = f(x) + xy + f(y)$ for every $x \in \mathbb{R}$ and every $y \in \{f(x) \mid x \in \mathbb{R}\}$, where $\mathbb{R}$ is the set of all real numbers.

15th Korean Mathematical Olympiad (4)

For $n \geq 3$, let $S = a_1 + a_2 + \cdots + a_n$ and $T = b_1 b_2 \cdots b_n$ for positive real numbers $a_1, a_2, \ldots, a_n, b_1, b_2, \ldots, b_n$, where the numbers b_i are pairwise distinct.

(a) Find the number of distinct real zeroes of the polynomial
$$f(x) = (x-b_1)(x-b_2)\cdots(x-b_n) \sum_{j=1}^{n} \frac{a_j}{x-b_j}.$$

(b) Prove the inequality
$$\frac{1}{n-1} \sum_{j=1}^{n} \left(1 - \frac{a_j}{S}\right) b_j > \left(\frac{T}{S} \sum_{j=1}^{n} \frac{a_j}{b_j}\right)^{\frac{1}{n-1}}.$$

15th Irish Mathematical Olympiad, (First Paper) (8)

Denote by $\mathbb{Q}$ the set of rational numbers. Determine all functions $f : \mathbb{Q} \to \mathbb{Q}$ such that
$$f(x + f(y)) = y + f(x), \quad \text{for all } x, y \in \mathbb{Q}.$$

Vietnamese Mathematical Olympiad, 2003 (3)

Find all polynomials $P(x)$ with real coefficients, satisfying the relation
$$(x^3 + 3x^2 + 3x + 2)P(x-1) = (x^3 - 3x^2 + 3x - 2)P(x)$$
for every real number x.

Vietnamese Mathematical Olympiad, 2003 (4)

Let $P(x) = 4x^3 - 2x^2 - 15x + 9$ and $Q(x) = 12x^3 + 6x^2 - 7x + 1$.

(i) Prove that each of these polynomials has three distinct real roots.

(ii) Let α and β be the greatest roots of $P(x)$ and $Q(x)$, respectively. Prove that $\alpha^2 + 3\beta^2 = 4$.

Vietnamese Mathematical Olympiad, 2003 (6)

Let f be a function defined on the set of real numbers $\mathbb{R}$, taking values in $\mathbb{R}$, and satisfying the condition $f(\cot x) = \sin 2x + \cos 2x$ for every x belonging to the open interval $(0, \pi)$. Find the least and the greatest values of the function $g(x) = f(x) \cdot f(1-x)$ on the closed interval $[-1, 1]$.

8th Macedonian Mathematical Competition (2)

Does there exist a function $f : \mathbb{N} \to \mathbb{N}$ such that for every $n \geq 2$,
$$f(f(n-1)) = f(n+1) - f(n) ?$$

British Mathematical Competition, 2003 (Round II) (4)

Let f be a function from the set of non-negative integers into itself such that, for all $n \geq 0$,

(i) $(f(2n+1))^2 - (f(2n))^2 = 6f(n) + 1$, and

(ii) $f(2n) \geq f(n)$.

How many numbers less than 2003 are there in the image of f?

Ukrainian Mathematical Olympiad, (11 Form) (6)

Find all functions $f : \mathbb{R} \to \mathbb{R}$ such that
$$f(xf(x) + f(y)) = x^2 + y$$
for all $x \in \mathbb{R}$ and $y \in \mathbb{R}$.

Belarus Mathematical Olympiad, 2003 (3)

Find all functions f from the real numbers to the real numbers such that, for any real numbers x and y,
$$f(xy)(f(x) - f(y)) = (x - y)f(x)f(y).$$

Indian TST for IMO, 2003 (3)

Find all functions $f : \mathbb{R} \to \mathbb{R}$ such that, for all x, y in $\mathbb{R}$, we have
$$f(x+y) + f(x)f(y) = f(x) + f(y) + f(xy).$$

Indian TST for IMO, 2003 (7)

Let $P(x)$ be a polynomial with integer coefficients such that $P(n) > n$ for all positive integers n. Suppose that for each positive integer m, there is a term in the sequence $P(1), P(P(1)), P(P(P(1))), \ldots$ which is divisible by m. Show that $P(x) = x + 1$.

Singapore Mathematical Olympiad, 2003 (4)

Find all real-valued functions $f : \mathbb{Q} \longrightarrow \mathbb{R}$ defined on the set of all rational numbers $\mathbb{Q}$ satisfying the conditions
$$f(x+y) = f(x) + f(y) + 2xy,$$
for all x, y in $\mathbb{Q}$ and $f(1) = 2002$. Justify your answers.

Iranian Mathematical Olympiad, 2002 (First Round) (5)

Let δ be a symbol such that $\delta \neq 0$ and $\delta^2 = 0$. Define
$$\begin{aligned}
\mathbb{R}[\delta] &= \{a + b\delta \mid a, b \in \mathbb{R}\} \\
a + b\delta = c + d\delta &\iff a = c \text{ and } b = d, \\
(a + b\delta) + (c + d\delta) &= (a + c) + (b + d)\delta, \\
(a + b\delta) \cdot (c + d\delta) &= ac + (ad + bc)\delta.
\end{aligned}$$

Let $P(x)$ be a polynomial with real coefficients. Show that $P(x)$ has a multiple root in $\mathbb{R}$ if and only if $P(x)$ has a non-real root in $\mathbb{R}[\delta]$.

Belarusian Mathematical Olympiad, 2002 (Category A) (2)

Let
$$P(x) = (x+1)^p(x-3)^q = x^n + a_1 x^{n-1} + a_2 x^{n-2} + \cdots + a_{n-1} x + a_n,$$
where p and q are positive integers.

(a) Given that $a_1 = a_2$, prove that $3n$ is a perfect square.

(b) Prove that there exist infinitely many pairs (p, q) of positive integers p and q such that the equality $a_1 = a_2$ is valid for the polynomial $P(x)$.

Belarusian Mathematical Olympiad, 2002 (Category A) (7)

Does there exist a surjective function $f : \mathbb{R} \to \mathbb{R}$ such that the expression $f(x+y) - f(x) - f(y)$ takes exactly two values 0 and 1 for various real x and y?

Singapore Mathematical Olympiad, 2003 (4)

Find all real-valued functions $f : \mathbb{Q} \longrightarrow \mathbb{R}$ defined on the set of all rational numbers $\mathbb{Q}$ satisfying the conditions

$$f(x+y) = f(x) + f(y) + 2xy,$$

for all x, y in $\mathbb{Q}$ and $f(1) = 2002$. Justify your answers.

Iranian Mathematical Olympiad, 2002 (First Round) (5)

Let δ be a symbol such that $\delta \neq 0$ and $\delta^2 = 0$. Define

$$\begin{aligned}
\mathbb{R}[\delta] &= \{a + b\delta \mid a, b \in \mathbb{R}\} \\
a + b\delta = c + d\delta &\iff a = c \text{ and } b = d, \\
(a + b\delta) + (c + d\delta) &= (a + c) + (b + d)\delta, \\
(a + b\delta) \cdot (c + d\delta) &= ac + (ad + bc)\delta.
\end{aligned}$$

Let $P(x)$ be a polynomial with real coefficients. Show that $P(x)$ has a multiple root in $\mathbb{R}$ if and only if $P(x)$ has a non-real root in $\mathbb{R}[\delta]$.

Belarusian Mathematical Olympiad, 2002 (Category A) (2)

Let

$$P(x) = (x+1)^p(x-3)^q = x^n + a_1 x^{n-1} + a_2 x^{n-2} + \cdots + a_{n-1} x + a_n,$$

where p and q are positive integers.

(a) Given that $a_1 = a_2$, prove that $3n$ is a perfect square.

(b) Prove that there exist infinitely many pairs (p, q) of positive integers p and q such that the equality $a_1 = a_2$ is valid for the polynomial $P(x)$.

Belarusian Mathematical Olympiad, 2002 (Category A) (7)

Does there exist a surjective function $f : \mathbb{R} \to \mathbb{R}$ such that the expression $f(x + y) - f(x) - f(y)$ takes exactly two values 0 and 1 for various real x and y?

34

f(u)nctions and p(o)lynomials
problems and <u>solutions</u>
from
Mathematical Olympiads

24-th Spanish Olympiad - First Round

Given the function f defined by $f(x) = \sqrt{4 + \sqrt{16x^2 - 8x^3 + x^4}}$.
(a) Draw the graph of the curve $y = f(x)$.
(b) Find, without the use of integral calculus, the area of the region bounded by the straight lines $x = 0$, $x = 6$, $y = 0$ and by the curve $y = f(x)$. Note: all the square roots are non-negative.

Solution

(a) Note that $f(x) = \sqrt{4 + |x^2 - 4x|}$. If $x > 4$ or $x < 0$ then $f(x) = |x - 2|$. If $0 < x < 4$ then $y = f(x)$ can be written as $(x-2)^2 + y^2 = (2\sqrt{2})^2$, a circle. Thus the graph is as shown.
(b) The area of $\triangle AOB$ and of $\triangle BCD$ is 2, the area of sector ABC is $\frac{1}{4}\pi(2\sqrt{2})^2 = 2\pi$, and the area of trapezoid $CDHE$ is $\frac{1}{2}(2+4) \times 2 = 6$. Thus the area of the region is $8 + 2\pi$.

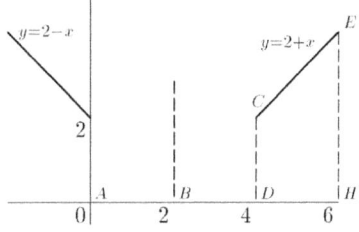

24-th Spanish Olympiad - First Round

Let $I_n = (n\pi - \pi/2, n\pi + \pi/2)$ and let f be the function defined by $f(x) = \tan x - x$.
(a) Show that the equation $f(x) = 0$ has only one root in each interval I_n, $n = 1, 2, 3, \ldots$.
(b) If c_n is the root of $f(x) = 0$ in I_n, find $\lim_{n \to \infty}(c_n - n\pi)$.

Solution

(a) This is obvious since $(\tan x - x)' = \sec^2 x - 1 \geq 0$ on I_n.
(b) As n goes to infinity the point of intersection of the line $y = x$ and the graph of $y = \tan x$ in the interval I_n goes off to infinity in each coordinate. Thus the difference $c_n - n\pi$ must go to $\pi/2$.

24-th Spanish Olympiad

Let f be a continuous function on $\mathbf{R}$ such that
 (i) $f(n) = 0$ for every integer n, and
 (ii) if $f(a) = 0$ and $f(b) = 0$ then $f(\frac{a+b}{2}) = 0$, with $a \neq b$.
Show that $f(x) = 0$ for all real x.

Solution

If n is an integer then $f(n/2) = 0$. By induction on m, we have $f(n/2^m) = 0$. Let s and ϵ be arbitrary real numbers. Since f is continuous at s, there is $\delta > 0$ such that $|f(s) - f(x)| < \epsilon$ whenever $|s - x| < \delta$. Since s has a 2-adic expansion, there is $n/2^m$ such that $|s - n/2^m| < \delta$. We now have $|f(s) - f(n/2^m)| = |f(s)| < \epsilon$. Hence $f(s) = 0$. (This is, of course, just the standard argument.)

Hungarian National Olympiad 1987

Determine the minimum of the function
$$f(x) = \sqrt{a^2 + x^2} + \sqrt{(b-x)^2 + c^2}$$
where a, b, c are positive numbers.

Solution

Let $AC = a$, $AB = b$ and $BD = c$. Let P be a point on AB and let $x = AP$, so that $BP = b - x$. Then $f(x) = CP + PD$. To minimize $CP + PD$, we follow the method of reflection (see Z. A. Melzak, *Companion to Mathematics*, John Wiley & Sons, 1973, pp. 26-27). Let D' be the reflection of D in the line AB. Since triangles PBD and PBD' are congruent, $PD = PD'$ and $f(x) = CP + PD'$. As x varies, P changes its position. But the distance $CP + PD'$ will be a minimum when P lies on the line CD'. The minimum value of $f(x)$ is then $CP + PD' = CD'$. Let CL be the perpendicular from C to the line DD'. From the right triangle CLD',

$$CD' = \sqrt{(LD)^2 + (CL)^2} = \sqrt{(a+c)^2 + b^2}.$$

Hungarian National Olympiad 1987

The domain of function f is $[0,1]$, and for any $x_1 \neq x_2$

$$|f(x_1) - f(x_2)| < |x_1 - x_2|.$$

Moreover, $f(0) = f(1) = 0$. Prove that for any x_1, x_2 in $[0,1]$,

$$|f(x_1) - f(x_2)| < \frac{1}{2}.$$

Solution

First $|f(x) - f(0)| \leq |x - 0|$, i.e. $|f(x)| \leq x$ and the inequality is strict for $x \neq 0$. Also $|f(x) - f(1)| \leq |x - 1|$, i.e. $|f(x)| \leq 1 - x$ with strict inequality for $x \neq 1$. Therefore

$$|f(x)| \leq \min(x, 1-x),$$

and the inequality is strict unless $x = 0$ or $x = 1$. Let $x_1, x_2 \in [0,1]$. If $|x_1 - x_2| \leq 1/2$, then

$$|f(x_1) - f(x_2)| \stackrel{*}{\leq} |x_1 - x_2| \leq 1/2.$$

This gives $|f(x_1) - f(x_2)| < 1/2$ since * is strict unless $x_1 = x_2$ and this case is trivial. So suppose $|x_1 - x_2| > 1/2$. Without loss of generality suppose that $x_1 \in (1/2, 1]$ and $x_2 \in [0, 1/2)$. Then

$$|f(x_1) - f(x_2)| \leq |f(x_1)| + |f(x_2)| \leq 1 - x_1 + x_2 = 1 - (x_1 - x_2) < \frac{1}{2}.$$

Therefore, in all cases $|f(x_1) - f(x_2)| < 1/2$ for all $x_1 \neq x_2$.

Romanian Mathematical Olympiad, Final Round 1978

Let P and Q be two polynomials (neither identically zero) with complex coefficients.
Show that P and Q have the same roots (with the same multiplicities) if and only if the function $f : \mathbf{C} \to \mathbf{R}$ defined by $f(z) = |P(z)| - |Q(z)|$ has a constant sign for all $z \in C$ if it is not identically zero.

Solution

The *only if* part is easy.

For the *if* part, we can assume without loss of generality that $f(z) \geq 0$. If r is any root of P, it immediately follows that it must also be a root of Q (note if P is a constant, then so also Q is a constant). Also the multiplicity of any root of P must be at most the corresponding multiplicity in Q. For if a root r had greater multiplicity in P than in Q, by setting $z = r + \varepsilon$, where ε is arbitrarily small, we would have $f(z) < 0$. Next the degree of P must be at least the degree of Q. Otherwise by taking $|z|$ arbitrarily large, we would have $f(z) < 0$. It follows that P and Q have the same roots with the same multiplicities.

Kozepiskolai Matematikai Lapok 1983

Does there exist a function $f : \mathbf{R} \to \mathbf{R}$ such that $\lim_{x \to \infty} f(x) = \infty$ and

$$\lim_{x \to \infty} \frac{f(x)}{\ln(\ln(\ldots(\ln x)\ldots))} = 0$$

holds for all n (where n is the number of logarithm functions in the denominator)?

Solution

The answer to the given problem is in the affirmative and it is a special case of a result of du Bois-Reymond [1].

First, if $f(x)/g(x) \to \infty$ as $x \to \infty$, we say that the order of f is greater than the order of g and we write it as $f \succ g$. The theorem of du Bois-Reymond is that given a scale of increasing functions φ_n such that

$$\varphi_1 \succ \varphi_2 \succ \varphi_3 \succ \ldots \succ 1,$$

then there exists an increasing function f such that $\varphi_n \succ f \succ 1$ for all values of n. Here we choose $\varphi_1 = \ln x$, $\varphi_2 = \ln \ln x$, $\varphi_3 = \ln \ln \ln x$, etc.

More generally we have the following: given a descending sequence $\{\varphi_n\}$: $\varphi_1 \succ \varphi_2 \succ \varphi_3 \succ \cdots \succ \varphi_n \succ \cdots \succ \varphi$ and an ascending sequence $\{\psi_n\}$: $\psi_1 \prec \psi_2 \prec \psi_3 \prec \cdots \prec \psi_p \prec \cdots \prec \psi$ such that $\psi_p \prec \varphi_n$ for all n and p then there is f such that $\psi_p \prec f \prec \varphi_n$ for all n and p.

Asian Pacific Mathematical Olympiad 1989

Determine all functions f from the reals to the reals for which
   (i) $f(x)$ is strictly increasing,
   (ii) $f(x) + g(x) = 2x$ for all real x where $g(x)$ is the composition inverse function to $f(x)$. (Note: f and g are said to be composition inverses if $f(g(x)) = x$ and $g(f(x)) = x$ for all real x.)

Solution

We will prove that $f(x) = x + d$ for some constant d, i.e., $f(x) - x$ is a constant function.

For each real d, denote by S_d the set of all x for which $f(x) - x = d$. Then we must show that exactly one S_d is nonempty. First we prove two lemmas.

Lemma 1. If $x \in S_d$ then $x + d \in S_d$.

Proof. Suppose $x \in S_d$. Then $f(x) = x+d$, so $g(x+d) = x$, and $f(x+d)+g(x+d) = 2x+2d$. Therefore $f(x+d) = x+2d$ and $x+d \in S_d$. □

Lemma 2. If $x \in S_d$ and $y \geq x$ then $y \notin S_{d'}$ for any $d' < d$.

Proof. First let y satisfy $x \leq y < x + (d-d')$. Note that by monotonicity $f(y) \geq f(x) = x+d$. Hence $y \in S_{d'}$ would imply $y+d' \geq x+d$ or $y \geq x+(d-d')$, a contradiction. Thus $y \notin S_{d'}$ in this case. Now by induction it follows that for all $x \in S_d$,

$$\text{if } x + (k-1)(d-d') \leq y < x + k(d-d') \text{ then } y \notin S_{d'}.$$

The base case $k = 1$ is proved above. Assume the statement holds for some k and let

$$x + k(d-d') \leq y < x + (k+1)(d-d').$$

Then

$$x + d + (k-1)(d-d') \leq y + d < x + d + k(d-d').$$

But $x+d \in S_d$, and so by the induction hypothesis $y + d' \notin S_{d'}$. The lemma follows. □

Now suppose that two S_d's are nonempty, say S_d and $S_{d'}$ where $d' < d$. If $0 < d'$, then $S_{d'}$ must contain arbitrarily large numbers by Lemma 1. But this is impossible by Lemma 2.

Editor's note. The above, slightly rewritten, is Evagelopoulos's argument, except he hadn't noted that his argument required $0 < d'$. We now complete the proof.

Lemma 3. If S_d and $S_{d'}$ are nonempty and $d' < d'' < d$ then $S_{d''}$ is also nonempty.

Proof. Since S_d and $S_{d'}$ are nonempty, there are x and x' so that $f(x) - x = d$ and $f(x') - x' = d'$. Since f is increasing and has an inverse, it is continuous, so the function $f(x) - x$ is also continuous. Thus by the Intermediate Value Theorem there is x'' so that $f(x'') - x'' = d''$, so $S_{d''} \neq \emptyset$. □

Now by Lemma 3 we need only consider two cases: $0 < d' < d$, which was handled by Evagelopoulos, and $d' < d < 0$. We do the second case. Choose some $y \in S_{d'}$. By Lemma 1, S_d contains arbitrarily large *negative* numbers, so we can find $x \in S_d$ such that $x < y$. But then $y \notin S_{d'}$ by Lemma 2. This contradiction completes the proof.

IMO Proposal by Greece (1989)

Let $g : \mathbf{C} \to \mathbf{C}$, $\omega \in \mathbf{C}$, $a \in \mathbf{C}$, with $\omega^3 = 1$ and $\omega \neq 1$. Show that there is one and only one function $f : \mathbf{C} \to \mathbf{C}$ such that
$$f(z) + f(\omega z + a) = g(z), \quad z \in \mathbf{C}.$$
Find the function f.

Solution

If $f : \mathbf{C} \to \mathbf{C}$ satisfies
$$g(z) = f(z) + f(\omega z + a) \quad \text{for all } z \in \mathbf{C}, \tag{1}$$
then
$$g(\omega z + a) = f(\omega z + a) + f(\omega^2 z + \omega a + a) \quad \text{for all } z \in \mathbf{C}. \tag{2}$$
Since $\omega^3 = 1$ and $1 + \omega + \omega^2 = 0$, we have
$$\omega^2 z + \omega a + a = \omega^2 z + a(1 + \omega) = \omega^2(z - a) = \frac{z - a}{\omega},$$
and so
$$g\left(\frac{z-a}{\omega}\right) = f\left(\frac{z-a}{\omega}\right) + f\left(\omega \cdot \frac{z-a}{\omega} + a\right) = f(\omega^2 z + \omega a + a) + f(z). \tag{3}$$
Therefore (1) minus (2) plus (3) yields
$$g(z) - g(\omega z + a) + g\left(\frac{z-a}{\omega}\right) = 2f(z),$$
and so
$$f(z) = \frac{1}{2}\left\{g(z) - g(\omega z + a) + g\left(\frac{z-a}{\omega}\right)\right\}.$$
On the other hand, it is easily checked that this choice of f satisfies (1) and hence it is the unique function with this property.

11th Austrian-Polish Mathematics Competition

Determine all strictly monotone increasing functions $f : \mathbf{R} \to \mathbf{R}$ satisfying the functional equation
$$f(f(x) + y) = f(x + y) + f(0)$$
for all $x, y \in \mathbf{R}$.

Solution

Let f be a strictly increasing function on $\mathbf{R}$ satisfying the functional equation. In particular, if we set $y = -x$, we must have $f(f(x) - x) = f(0) + f(0) = 2f(0)$. Unless $f(x) - x$ is a constant we have a contradiction to the fact that f is $1-1$ since $f(f(x) - x)$ is constant. Thus $f(x) = x + c$ where $c = f(0)$. Any such f is strictly increasing, and we have
$$f(f(x) + y) = f(x + c + y) = (x + y + c) + c = f(x + y) + f(0),$$
that is, any such f is a solution.

Australian Mathematical Olympiad 1985

Find all polynomials $f(x)$ with real coefficients such that
$$f(x) \cdot f(x+1) = f(x^2 + x + 1).$$

Solution

Substituting $x - 1$ for x in the original equation
$$f(x)f(x+1) = f(x^2 + x + 1) \tag{1}$$
we get
$$f(x-1) \cdot f(x) = f(x^2 - x + 1). \tag{2}$$
There are two cases.

Case 1. If $f(x)$ is a constant polynomial we have $f(x) \equiv 0$ or $f(x) \equiv 1$.

Case 2. Suppose $f(x)$ is not a constant. Then $f(x)$ has at least one (complex) root. Let a be a root with maximum absolute value. By (1) and (2), $f(a) = 0$ implies $f(a^2 + a + 1) = 0$, and $f(a^2 - a + 1) = 0$. Thus $a \neq 0$. If $a^2 + 1 \neq 0$, then a, $a^2 + a + 1$, $a^2 - a + 1$, $-a$ are the vertices of a parallelogram, and $|a^2 + a + 1|$ or $|a^2 - a + 1|$ is greater than $|a|$, contradicting its choice. So $a = \pm i$, and since f has real coefficients *both i and $-i$* are roots of $f(x)$ and $f(x) = (x^2 + 1)^m g(x)$ where m is a positive integer, and $g(x)$ is a polynomial which has real coefficients and is not divisible by $x^2 + 1$. By (1)

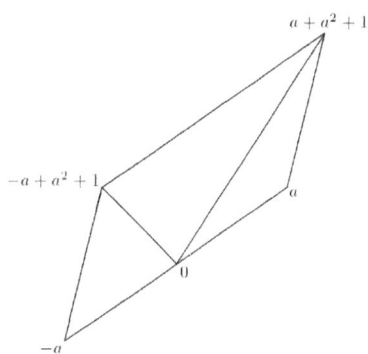

$$(x^2 + 1)^m g(x) \cdot (x^2 + 2x + 2)^m g(x+1) = (x^4 + 2x^3 + 3x^2 + 2x + 2)^m g(x^2 + x + 1).$$

Now
$$(x^2 + 1)(x^2 + 2x + 2) = x^4 + 2x^3 + 3x^2 + 2x + 2. \tag{3}$$
This gives that
$$g(x) \cdot g(x+1) = g(x^2 + x + 1),$$
i.e. $g(x)$ satisfies the same functional equation as $f(x)$. By the argument at the beginning of this case, $g(x)$ (not being divisible by $x^2 + 1$) must be a constant polynomial, and hence, by Case 1, $g(x) \equiv 1$. Thus if $f(x)$ is non-constant and satisfies (1) we must have that
$$f(x) = (x^2 + 1)^m.$$
On the other hand (3) shows that (1) is then satisfied.

Singapore MSI Mathematical Competition 1988

Let $f(x)$ be a polynomial of degree n such that $f(k) = \frac{k}{k+1}$ for each $k = 0, 1, 2, \ldots, n$. Find $f(n+1)$.

Solution

Let $g(x) = (x+1)f(x) - x$. Then the given condition becomes $g(0) = g(1) = \cdots = g(n) = 0$. It follows that $g(x) = kx(x-1)\cdots(x-n)$ and

$$(x+1)f(x) = x + kx(x-1)\cdots(x-n).$$

Putting $x = -1$, we have $k = \frac{(-1)^{n+1}}{(n+1)!}$. We conclude that

$$f(x) = \frac{x + \frac{(-1)^{n+1}}{(n+1)!}x(x-1)\cdots(x-n)}{x+1}.$$

Thus

$$f(n+1) = \frac{n+1+(-1)^{n+1}}{n+2} = \begin{cases} 1 & n \text{ odd} \\ \frac{n}{n+2} & n \text{ even.} \end{cases}$$

First Selection Test of the Chinese I.M.O. Team 1988

Determine all functions f from the rational numbers to the complex numbers such that
(i) $f(x_1 + x_2 + \cdots + x_{1988}) = f(x_1)f(x_2)\ldots f(x_{1988})$
for all rational numbers $x_1, x_2, \ldots, x_{1988}$, and
(ii) $\overline{f(1988)}f(x) = f(1988)\overline{f(x)}$
for all rational numbers x, where $\overline{z}$ denotes the complex conjugate of z.

Solution

Suppose $f \not\equiv 0$. Since $f(x) = f(x)f(0)^{1987}$, we have $(f(0))^{1987} = 1$. Note that $f(x+y) = f(x)f(y)(f(0))^{1986}$. Let $g(x) = f(x)/f(0)$. Then

$$g(x+y) = \frac{f(x+y)}{f(0)} = \frac{f(x)f(y)(f(0))^{1986}}{f(0)} = \frac{f(x)}{f(0)} \cdot \frac{f(y)}{f(0)} \cdot (f(0))^{1987} = g(x)g(y).$$

It follows easily that $g(x) = e^{bx}$ for some complex number b. From this we get that $f(x) = f(0)e^{bx}$. Since the condition (ii) implies $e^{(b-\overline{b})(x-1988)} = 1$ we obtain that b must be real. Answer: $f \equiv 0$ or $f(x) = ae^{bx}$ where $a^{1987} = 1$ and b is real.

43rd Mathematical Olympiad (1991-92) in Poland (Final round)

Determine all functions f defined on the set of positive rational numbers, taking values in the same set, which satisfy for every positive rational number x the conditions

$$f(x+1) = f(x) + 1 \quad \text{and} \quad f(x^3) = (f(x))^3.$$

Solution

Let $\mathbb{N}$ and $\mathbb{Q}^+$ denote the set of positive integers, and the set of positive rational numbers, respectively. We show that $f(x) = x$, for all $x \in \mathbb{Q}^+$, is the only function satisfying the given conditions. First of all, by the first condition and an easy induction we see that $f(x+n) = f(x) + n$, for all $x \in \mathbb{Q}^+$, and for all $n \in \mathbb{N}$. Now for arbitrary $\frac{p}{q} \in \mathbb{Q}^+$, where $p, q \in \mathbb{N}$, we have

$$f\left(\left(\frac{p}{q} = q^2\right)^3\right) = f\left(\frac{p^3}{q^3} + 3p^2 + 3pq^3 + q^6\right)$$
$$= f\left(\left(\frac{p}{q}\right)^3\right) + 3p^2 + 3pq^3 + q^6. \quad (1)$$

On the other hand

$$f\left(\left(\frac{p}{q} + q^2\right)^3\right) = f\left(\left(\frac{p}{q} + q^2\right)\right)^3 = \left(f\left(\frac{p}{q}\right) + q^2\right)^3$$
$$= \left(f\left(\frac{p}{q}\right)\right)^3 + 3\left(f\left(\frac{p}{q}\right)\right)^2 q^2 + 3\left(f\left(\frac{p}{q}\right)\right)^2 \cdot q^4 + q^6. \quad (2)$$

Letting $t = f(\frac{p}{q})$ and comparing (1) and (2), we get, since $f(\frac{p^3}{q^3}) = (f(\frac{p}{q}))^3$, $p(p+q^3) = q^2 t^2 + q^4 t$ or $q^2 t^2 + q^4 t - p(p+q^3) = 0$ or $(qt-p)(qt+p+q^3) = 0$. Since $qt + p + q^3 > 0$, we must have $t = \frac{p}{q}$, i.e. $f(\frac{p}{q}) = \frac{p}{q}$, and we are done.

43rd Mathematical Olympiad (1991-92) in Poland (Final round) — 4

Define the sequence of functions $f_0, f_1, f_2, \ldots$ by

$$f_0(x) = 8 \quad \text{for all } x \in \mathbb{R},$$
$$f_{n+1}(x) = \sqrt{x^2 + 6f_n(x)} \quad \text{for } n = 0, 1, 2, \ldots \text{ and for all } x \in \mathbb{R}.$$

For every positive integer n, solve the equation $f_n(x) = 2x$.

Solution

Since $f_n(x)$ is positive, $f_n(x) = 2x$ has only positive solutions. We show that, for each n, $f_n(x) = 2x$ has a solution $x = 4$. Since $f_1(x) = \sqrt{x^2 + 48}$, $x = 4$ is a solution of $f_2(x) = 2x$. Now $f_{n+1}(4) = \sqrt{4^2 + 6f_n(4)} = \sqrt{4^2 + 6 \cdot 8} = 8 = 2 \cdot 4$, which completes the inductive step. Next, induction on n gives us that for each n, $\frac{f_n(x)}{x}$ decreases as x increases in $(0, \infty)$. It follows that $f_n(x) = 2x$ has the unique solution $x = 4$.

Canadian Mathematical Olympiad 1996

We denote an arbitrary permutation of the integers $1, \ldots, n$ by $a_1, \ldots, a_n$. Let $f(n)$ be the number of these permutations such that
(i) $a_1 = 1$;
(ii) $|a_i - a_{i+1}| \leq 2$, $i = 1, \ldots, n-1$.
Determine whether $f(1996)$ is divisible by 3.

Solution

Let $a_1, a_2, \ldots, a_n$ be a permutation of $1, 2, \ldots, n$ with properties (i) and (ii).

A crucial observation, needed in Case II (b) is the following: If a_k and a_{k+1} are consecutive integers (i.e. $a_{k+1} = a_k \pm 1$), then the terms to the right of a_{k+1} (also to the left of a_k) are either all less than both a_k and a_{k+1} or all greater than both a_k and a_{k+1}.

Since $a_1 = 1$, by (ii) a_2 is either 2 or 3.

CASE I: Suppose $a_2 = 2$. Then $a_3, a_4, \ldots, a_n$ is a permutation of $3, 4, \ldots, n$. Thus $a_2, a_3, \ldots, a_n$ is a permutation of $2, 3, \ldots, n$ with $a_2 = 2$ and property (ii). Clearly there are $f(n-1)$ such permutations.

CASE II: Suppose $a_2 = 3$.
(a) Suppose $a_3 = 2$. Then $a_4, a_5, \ldots, a_n$ is a permutation of $4, 5, \ldots, n$ with $a_4 = 4$ and property (ii). There are $f(n-3)$ such permutations.
(b) Suppose $a_3 \geq 4$. If a_{k+1} is the first even number in the permutation then, because of (ii), $a_1, a_2, \ldots, a_k$ must be $1, 3, 5, \ldots, 2k-1$ (in that order). Then a_{k+1} is either $2k$ or $2k-2$, so that a_k and a_{k+1} are consecutive integers. Applying the crucial observation made above, we deduce that $a_{k+2}, \ldots, a_n$ are all either greater than or smaller than a_k and a_{k+1}. But 2 must be to the right of a_{k+1}. Hence $a_{k+2}, \ldots, a_n$ are the even integers less than a_{k+1}. The only possibility then, is

$$1, 3, 5, \ldots, a_{k-1}, a_k, \ldots, 6, 4, 2.$$

Cases I and II show that
$$f(n) = f(n-1) + f(n-3) + 1, \quad n \geq 4. \tag{$\star$}$$
Calculating the first few values of $f(n)$ directly gives

$$f(1) = 1, \ f(2) = 1, \ f(3) = 2, \ f(4) = 4, \ f(5) = 6.$$

Calculating a few more $f(n)$'s using $(\star)$ and mod 3 arithmetic, $f(1) = 1$, $f(2) = 1$, $f(3) = 2$, $f(4) = 1$, $f(5) = 0$, $f(6) = 0$, $f(7) = 2$, $f(8) = 0$, $f(9) = 1$, $f(10) = 1$, $f(11) = 2$. Since $f(1) = f(9)$, $f(2) = f(10)$ and $f(3) = f(11)$ mod 3, $(\star)$ shows that

$$f(a) = f(a \bmod 8), \bmod 3, \ a \geq 1.$$

Hence $f(1996) \equiv f(4) \equiv 1 \pmod{3}$ so 3 does not divide $f(1996)$.

Canadian Mathematical Olympiad 1996

Let $r_1, r_2, \ldots, r_m$ be a given set of m positive rational numbers such that $\sum_{k=1}^{m} r_k = 1$. Define the function f by $f(n) = n - \sum_{k=1}^{m}[r_k n]$ for each positive integer n. Determine the minimum and maximum values of $f(n)$. Here $[x]$ denotes the greatest integer less than or equal to x.

Solution

Let
$$f(n) = n - \sum_{k=1}^{m}[r_k n]$$
$$= n\sum_{k=1}^{m} r_k - \sum_{k=1}^{m}[r_k n]$$
$$= \sum_{k=1}^{m}\{r_k n - [r_k n]\}.$$

Now $0 \leq x - [x] < 1$, and if c is an integer, $(c + x) - [c + x] = x - [x]$.

Hence $0 \leq f(n) < \sum_{k=1}^{m} 1 = m$. Because $f(n)$ is an integer, $0 \leq f(n) \leq m - 1$.

To show that $f(n)$ can achieve these bounds for $n > 0$, we assume that $r_k = \dfrac{a_k}{b_k}$ where a_k, b_k are integers; $a_k < b_k$.

Then, if $n = b_1 b_2 \ldots b_m$, $(r_k n) - [r_k n] = 0$, $k = 1, 2, \ldots, m$ and thus $f(n) = 0$.

Letting $n = b_1 b_2 \ldots b_n - 1$, then
$$\begin{aligned} r_k n &= r_k(b_1 b_2 \ldots b_m - 1) \\ &= r_k\{(b_1 b_2 \ldots b_m - b_k) + b_k - 1)\} \\ &= \text{integer} + r_k(b_k - 1). \end{aligned}$$

This gives
$$\begin{aligned} r_k n - [r_k n] &= r_k(b_k - 1) - [r_k(b_k - 1)] \\ &= \frac{a_k}{b_k}(b_k - 1) - \left[\frac{a_k}{b_k}(b_k - 1)\right] \\ &= \left(a_k - \frac{a_k}{b_k}\right) - \left[a_k - \frac{a_k}{b_k}\right] \\ &= \left(a_k - \frac{a_k}{b_k}\right) - (a_k - 1) \\ &= 1 - \frac{a_k}{b_k} = 1 - r_k. \end{aligned}$$

Hence
$$f(n) = \sum_{k=1}^{m}(1 - r_k) = m - 1.$$

10th Iranian Mathematical Olympiad (Second Stage Exam)

Let X be a non-empty finite set and $f : X \to X$ a function such that for all x in X, $f^p(x) = x$, where p is a constant prime. If $Y = \{x \in X : f(x) \neq x\}$, prove that the number of elements of Y is divisible by p.

Solution

For each $x \in Y$ consider that *orbit*, $B(x)$, of x defined by
$$B(x) = \{x, f(x), f^2(x), \ldots, f^{p-1}(x)\}.$$
We claim that all the elements of $B(x)$ are distinct.

Suppose not. Then let j be the *least* positive integer such that $f^i(x) = f^j(x)$ for some integer i with $0 \leq i < j \leq p-1$. (We define $f^i(x) = x$ if $i = 0$.) Then
$$x = f^p(x) = f^{p-j}(f^j(x)) = f^{p-j}(f^i(x)) = f^{p-j+i}(x)$$
$$\Rightarrow f^{j-i}(x) = f^{j-i}(f^{p-j+i}(x)) = f^p(x) = x.$$
Since $0 < j - i \leq j$, we must have $i = 0$ and thus $f^j(x) = x$. Now let $p = qj + r$, where q, r are integers with $q > 0$ and $0 \leq r < j$. Clearly $f^j(x) = x$ implies $f^{qj}(x) = x$ and hence
$$f^r(x) = f^r(f^{qj}(x)) = f^p(x) = x.$$
Since $r < j$, we must have $r = 0$ and thus $p = qj$. Since $f(x) \neq x$, $j > 1$. On the other hand, since $j < p$, $q > 1$. Hence p is a composite, a contradiction. Therefore, $f^i(x) \neq f^j(x)$ for all $i = 0, 1, 2, \ldots, p-1$, we see that $B(x) \subset Y$.

Next we show that the orbits of two elements of Y are either *disjoint* or *identical*. Let $x, y \in Y$ and suppose $B(x) \cap B(y) \neq \emptyset$. Then $f^l(x) = f^k(y)$ for some integers l and k, with $0 \leq l \leq k \leq p-1$. Hence
$$y = f^k(y) = f^{p-k}(f^k(y)) = f^{p-k}(f^l(x)) = f^{p-k+l}(x),$$
which show that $y \in B(x)$. It then follows that $B(x) = B(y)$. Therefore Y can be partitioned into disjoint orbits each having cardinality p and the result follows.

Remarks. (1) Actually, the result still holds even when $p = 1$ since in this case $Y = \emptyset$ and thus $|Y| = 0$. (2) The result need not hold if p is composite. A counterexample is given by $X = \{1, 2, 3, 4, 5, 6\}$, $f(1) = 2$, $f(2) = 1$, $f(3) = 4$, $f(4) = 5$, $f(5) = 6$, and $f(6) = 3$. In this case, $p = 4$ and $Y = X$, $|Y| = 6$.

6th Korean Mathematical Olympiad (Final Round), 1993

Let n be a given natural number. Find all the continuous functions $f(x)$ satisfying:

$$\binom{n}{0}f(x) + \binom{n}{1}f(x^2) + \binom{n}{2}f(x^{2^2}) + \cdots + \binom{n}{n-1}f(x^{2^{n-1}}) + \binom{n}{n}f(x^{2^n})$$

Solution

We prove by induction that $f(x) \equiv 0$ if f is a continuous function satisfying the condition (*). With $n = 0$ there is nothing to prove as the condition becomes $f(x) = 0$ for all x. For $n = 1$, assume that $f(x) + f(x^2) = 0$ for all x with f continuous. Now $f(-x) = -f(x^2) = f(x)$ so f is an even function and it suffices to prove the result for $x \geq 0$. From $f(0) = -f(0)$ and $f(1) = -f(1)$ we get $f(0) = 0 = f(1)$. So consider a fixed value of $x > 0$. Now $\lim_{k \to \infty} x^{1/k} = 1$ from which we obtain $\lim_{m \to \infty} x^{1/2^m} = 0$. It is easy to see that $f(x^{1/2^m}) = (-1)^m f(x)$ for $m \geq 0$. However, $\lim_{m \to \infty} f(x^{1/2^m}) = f(1) = 0$. From this it follows that $f(x) = 0$, establishing the result for $n = 1$.

Now assume f is a continuous function satisfying

$$\sum_{j=0}^{n+1} \binom{n+1}{j} f(x^{2^j}) = 0 \quad \text{for all } x.$$

Set

$$g(x) = \sum_{l=0}^{n} \binom{n}{l} f(x^{2^l}).$$

Now

$$\begin{aligned}
g(x) + g(x^2) &= \sum_{l=0}^{n} \binom{n}{l} f(x^{2^l}) + \sum_{l=0}^{n} \binom{n}{l} f((x^2)^{2^l}) \\
&= \sum_{l=0}^{n} \binom{n}{l} f(x^{2^l}) + \sum_{l=0}^{n} \binom{n}{l} f(x^{2^{l+1}}) \\
&= \sum_{j=0}^{n+1} \binom{n+1}{j} f(x^{2^j}) = 0
\end{aligned}$$

since $\binom{n+1}{j} = \binom{n}{j} + \binom{n}{j-1}$.

But then $g(x)$ is a continuous function satisfying $g(x) + g(x^2) = 0$ for all x. By the case $n = 1$, $g(x) = 0$ for all x, i.e.

$$g(x) = \sum_{l=0}^{n} \binom{n}{l} f(x^{2^l}) = 0 \text{ for all } x.$$

By the induction hypothesis $f(x) = 0$ for all x and the induction is complete.

6th Irish Mathematical Olympiad, 1993

Let $a_0, a_1, \ldots, a_{n-1}$ be real numbers, where $n \geq 1$, and let $f(x) = x^n + a_{n-1}x^{n-1} + \cdots + a_0$ be such that $|f(0)| = f(1)$ and each root α of f is real and satisfies $0 < \alpha < 1$. Prove that the product of the roots does not exceed $1/2^n$.

Solution

Let $f(x) = (x - \alpha_1)(x - \alpha_2)\ldots(x - \alpha_n)$ where the α_i denote the n real roots of f, $i = 1, 2, \ldots, n$. Then from $|f(0)| = f(1)$ we get $\prod(1 - \alpha_i) = \prod \alpha_i$. (All products are over $i = 1, 2, \ldots, n$.) Using the Arithmetic Mean–Geometric Mean Inequality we then get

$$\left(\prod \alpha_i\right)^2 = \prod \alpha_i(1 - \alpha_i) \leq \prod \left(\frac{\alpha_i + (1 - \alpha_i)}{2}\right)^2 = \frac{1}{2^{2n}}$$

from which $\prod \alpha_i \leq \frac{1}{2^n}$ follows. Equality holds if and only if $\alpha_i = \frac{1}{2}$ for all $i = 1, 2, \ldots, n$.

Turkish Mathematical Olympiad, 1993 (Final Selection Test)

Let $\mathbb{Q}^+$ denote the set of all positive rational numbers. Find all functions $f : \mathbb{Q}^+ \to \mathbb{Q}^+$ such that

for every $x, y \in \mathbb{Q}^+$, $\quad f\left(x + \dfrac{y}{x}\right) = f(x) + \dfrac{f(y)}{f(x)} + 2y.$

Solution

We show that the only solution is $f(x) = x^2$, $x \in \mathbb{Q}^+$.
Take $x = 1$, this gives

$$f(1 + y) = f(1) + \frac{f(y)}{f(1)} + 2y. \tag{1}$$

Now taking $x = y$ we get

$$f(y + 1) = f(y) + 1 + 2y. \tag{2}$$

From (1), and (2) we obtain

$$(f(1) - 1)\left(1 - \frac{f(y)}{f(1)}\right) = 0.$$

Now f can **not** be a constant function, so we have that

$$f(1) = 1. \tag{3}$$

By induction, from (2) and (3) we get

$$f(k) = k^2 \quad k \in \mathbb{N}^+. \tag{4}$$

Therefore, if $n, k \in \mathbb{N}^+$, using (4), we obtain that

$$\begin{aligned} f\left(k + \frac{n}{k}\right) &= f(k) + \frac{f(n)}{f(k)} + 2n \\ &= k^2 + \frac{n^2}{k^2} + 2n = \left(k + \frac{n}{k}\right)^2. \end{aligned} \tag{5}$$

Let $x = \frac{m}{k} \in \mathbb{Q}^+$, and $N \in \mathbb{N}^+$. By (2), we have

$$\begin{aligned} f(x + N) - f(x) &= \sum_{j=0}^{N-1}[f(x + j + 1) - f(x + j)] \\ &= \sum_{j=0}^{N-1}[1 + 2(x + j)] \\ &= N(2x + 1) + (N - 1)N \\ &= (x + N)^2 - x^2. \end{aligned}$$

Therefore $f(x + N) - (x + N)^2 = f(x) - x^2$, $N \in \mathbb{N}^+$, $x \in \mathbb{Q}^+$.
In particular, using (5), we obtain that

$$\begin{aligned} 0 = f\left(\frac{n}{k} + k\right) - \left(\frac{n}{k} + k\right)^2 &= f\left(\frac{n}{k}\right) - \left(\frac{n}{k}\right)^2 \\ &= f(x) - x^2, \end{aligned}$$

where $x = \frac{n}{k}$, with $n, k \in \mathbb{N}^+$.
Thus $f(x) = x^2$ for $x \in \mathbb{Q}^+$.

Canadian Mathematical Olympiad, 1997

Write the sum

$$\sum_{k=0}^{n} \frac{(-1)^k \binom{n}{k}}{k^3 + 9k^2 + 26k + 24},$$

in the form $\frac{p(n)}{q(n)}$, where p and q are polynomials with integer coefficients.

Solution

We first note that

$$k^3 + 9k^2 + 26k + 24 = (k+2)(k+3)(k+4).$$

Let $S(n) = \sum_{k=0}^{n} \frac{(-1)^k \binom{n}{k}}{k^2 + 9k^2 + 26k + 24}$.

Then

$$\begin{aligned} S(n) &= \sum_{k=0}^{n} \frac{(-1)^k n!}{k!(n-k)!(k+2)(k+3)(k+4)} \\ &= \sum_{k=0}^{n} \left(\frac{(-1)^k (n+4)!}{(k+4)!(n-k)!} \right) \times \left(\frac{k+1}{(n+1)(n+2)(n+3)(n+4)} \right). \end{aligned}$$

Let

$$T(n) = (n+1)(n+2)(n+3)(n+4)S(n) = \sum_{k=0}^{n} \left((-1)^k \binom{n+4}{k+4}(k+1) \right).$$

Now, for $n \geq 1$,

$$\sum_{i=0}^{n} (-1)^i \binom{n}{i} = 0 \qquad (*)$$

since

$$(1-1)^n = \binom{n}{0} - \binom{n}{1} + \binom{n}{2} + \ldots + (-1)^n \binom{n}{n} = 0.$$

Also

$$\sum_{i=0}^{n}(-1)^{i}\binom{n}{i}i = \sum_{i=1}^{n}(-1)^{i}\frac{i \cdot n!}{i! \cdot (n-i)!} + (-1)^{0} \cdot \frac{0 \cdot n!}{0! \cdot n!}$$

$$= \sum_{i=1}^{n}(-1)^{i}\frac{n!}{(i-1)!(n-i)!}$$

$$= \sum_{i=1}^{n}(-1)^{i}n\binom{n-1}{i-1}$$

$$= n\sum_{i=1}^{n}(-1)^{i}\binom{n-1}{i-1}$$

$$= -n\sum_{i=1}^{n}(-1)^{i-1}\binom{n-1}{i-1}.$$

Substituting $j = i - 1$, (*) shows that

$$\sum_{i=0}^{n}(-1)^{i}\binom{n}{i}i = -n\sum_{j=0}^{n-1}(-1)^{j}\binom{n-1}{j} = 0. \qquad (**)$$

Hence

$$T(n) = \sum_{k=0}^{n}(-1)^{k}\binom{n+4}{k+4}(k+1)$$

$$= \sum_{k=0}^{n}(-1)^{k+4}\binom{n+4}{k+4}(k+1)$$

$$= \sum_{k=-4}^{n}(-1)^{k+4}\binom{n+4}{k+4}(k+1)$$

$$- \left(-3 + 2(n+4) - \binom{n+4}{2}\right).$$

Substituting $j = k + 4$, gives

$$T(n) = \sum_{j=0}^{n+4}(-1)^{j}\binom{n+4}{j}(j-3) - \left(2n + 8 - 3 - \frac{(n+4)(n+3)}{2}\right)$$

$$= \sum_{j=0}^{n+4}(-1)^{j}\binom{n+4}{j}j$$

$$-3\sum_{j=0}^{n+4}(-1)^{j}\binom{n+4}{j} - \frac{1}{2}(4n + 10 - n^{2} - 7n - 12)$$

The first two terms are zero because of results (*) and (**) so

$$T(n) = \frac{n^2 + 3n + 2}{2}.$$

Then
$$\begin{aligned}
S(n) &= \frac{T(n)}{(n+1)(n+2)(n+3)(n+4)} \\
&= \frac{n^2 + 3n + 2}{2(n+1)(n+2)(n+3)(n+4)} \\
&= \frac{(n+1)(n+2)}{2(n+1)(n+2)(n+3)(n+4)} \\
&= \frac{1}{2(n+3)(n+4)}.
\end{aligned}$$

Therefore
$$\sum_{k=0}^{n} \frac{(-1)^k \binom{n}{k}}{k^3 + 9k^2 + 26k + 24} = \frac{1}{2(n+3)(n+4)}.$$

16th Austrian Polish Mathematical Competition

Let the function f be defined as follows:

If $n = p^k > 1$ is a power of a prime number p, then $f(n) := n + 1$.

If $n = p_1^{k_1} \cdots p_r^{k_r}$ ($r > 1$) is a product of powers of pairwise different prime numbers, then $f(n) := p_1^{k_1} + \cdots + p_r^{k_r}$.

For every $m > 1$ we construct the sequence $\{a_0, a_1, \ldots\}$ such that $a_0 = m$ and $a_{j+1} = f(a_j)$ for $j \geq 0$. We denote by $g(m)$ the smallest element of this sequence. Determine the value of $g(m)$ for all $m > 1$.

Solution

It is evident that $g(2) = 2$, $g(3) = 3$, $g(4) = 4$, $g(5) = 5$, $g(6) = 6$, $g(7) = 7$.

Lemma. For $m > 6$, $g(m) > 6$.

Proof. Consider $m > 6$, and its associated sequence.

$$6 = 1 + 5 = 2 + 4.$$

Now $2 + 4$ is not permissible since $(2, 4) \neq 1$, and the only way 6 can be part of the sequence is if 5 occurs earlier.

$$5 = 1 + 4 = 2 + 3.$$

So 5 is only part of the sequence if 4 or 6 occur earlier.

$$4 = 1 + 3 = 2 + 2,$$

so 4 is part of the sequence only if 3 occurs earlier.

$$3 = 1 + 2,$$

so 3 is part of the sequence only if 2 occurs earlier, which is impossible with $a_0 > 6$.

Consider the next several values of $g(m)$. By calculation $g(8) = 7 = g(9) = g(10) = g(11) = g(12) = g(13) = g(14) = g(15) = g(16)$. We next prove by induction that if $m \geq 16$ then $g(m) = 7$. We may assume $m > 16$.

Observe that at least one of $m, m+1, m+2, m+3, m+4, m+5$ is not a power of a prime because it is congruent to $0 \bmod 6$. Letting n be the first value which is not a power of a prime $m \leq n \leq m + 5$. The result follows immediately from the following since $g(m) \leq g(n) \leq f(n)$.

Lemma. Suppose $n > 16$ is not a power of a prime. Then $f(n) < n - 5$.

Proof. Write $n = ab$ with $1 < a < b$ and $(a, b) = 1$. Then $b \geq 5$. Now

$$a + b < ab - 5$$

is equivalent to $1 + \frac{a}{b} + \frac{5}{b} < a$ which is immediate unless $a = 2$ because $\frac{a}{b} < 1$, $\frac{5}{b} \leq 1$. But if $a = 2$ then $b > 8$ as $ab > 16$ and

$$1 + \frac{a}{b} + \frac{5}{b} < 1 + \frac{2}{8} + \frac{5}{8} < 2.$$

The weaker inequality $a + b < ab$ for $1 < a < b$ is immediate. It follows that

$$f(n) < a + b < ab - 5$$

if $n > 16$.

This completes the proof.

Romanian First Team Selection Test, 1993 (34th IMO)

Find the greatest real number a such that
$$\frac{x}{\sqrt{y^2+z^2}} + \frac{y}{\sqrt{z^2+x^2}} + \frac{z}{\sqrt{x^2+y^2}} > a$$
is true for all positive real numbers x, y, z.

Solution

We claim that $a = 2$. Let
$$f(x,y,z) = \frac{x}{\sqrt{y^2+z^2}} + \frac{y}{\sqrt{z^2+x^2}} + \frac{z}{\sqrt{x^2+y^2}}.$$

We show that $f(x,y,z) > 2$. Since $f(x,y,z) \to 2$ as $x \to y$ and $z \to 0$, the lower bound 2 is sharp. Without loss of generality, assume that $x \geq y \geq z$. Since by the arithmetic–harmonic-mean inequality, we have
$$\frac{\sqrt{z^2+x^2}}{\sqrt{y^2+z^2}} + \frac{\sqrt{y^2+z^2}}{\sqrt{z^2+x^2}} \geq 2,$$

it suffices to show that
$$f(x,y,z) > \frac{\sqrt{z^2+x^2}}{\sqrt{y^2+z^2}} + \frac{\sqrt{y^2+z^2}}{\sqrt{z^2+x^2}}$$

or equivalently,
$$\frac{z}{\sqrt{x^2+y^2}} > \frac{\sqrt{z^2+x^2}-x}{\sqrt{y^2+z^2}} + \frac{\sqrt{y^2+z^2}-y}{\sqrt{z^2+x^2}}.$$

By simple algebra, this is easily seen to be equivalent to
$$\frac{z}{\sqrt{y^2+z^2}(\sqrt{z^2+x^2}+x)} + \frac{z}{\sqrt{z^2+x^2}(\sqrt{y^2+z^2}+y)} < \frac{1}{\sqrt{x^2+y^2}}. \quad (1)$$

Since $\sqrt{y^2+z^2} \geq \sqrt{2z^2} = \sqrt{2}\,z$, $\sqrt{z^2+x^2} > x$ and $\sqrt{2}\,x \geq \sqrt{x^2+y^2}$, we have
$$\frac{z}{\sqrt{y^2+z^2}(\sqrt{z^2+x^2}+x)} < \frac{z}{\sqrt{2}\,z(x+x)} = \frac{1}{2\sqrt{2}\,x} \leq \frac{1}{2\sqrt{x^2+y^2}}.$$

Thus to establish (1), it remains to show that
$$\frac{z}{\sqrt{z^2+x^2}(\sqrt{y^2+z^2}+y)} < \frac{1}{2\sqrt{x^2+y^2}}$$

or equivalently
$$\frac{2z}{\sqrt{y^2+z^2}+y} < \sqrt{\frac{z^2+x^2}{x^2+y^2}}.$$

Since
$$\frac{z^2+x^2}{x^2+y^2} = 1 - \frac{y^2-z^2}{x^2+y^2},$$
which is an non-decreasing function of x, we have
$$\frac{z^2+x^2}{x^2+y^2} \geq \frac{z^2+y^2}{2y^2},$$
and thus it suffices to show that
$$\frac{\sqrt{z^2+y^2}}{\sqrt{2}\, y} > \frac{2z}{\sqrt{y^2+z^2}+y},$$
or equivalently
$$y^2+z^2+y\sqrt{y^2+z^2} > 2\sqrt{2}\, yz. \qquad (2)$$
Since $y^2+z^2 \geq 2yz$, we have
$$\begin{aligned} y^2+z^2+y\sqrt{y^2+z^2} &\geq 2yz + y\sqrt{2z^2} \\ &= (2+\sqrt{2})yz > 2\sqrt{2}\, yz \end{aligned}$$
and thus (2) holds. This completes the proof.

Romanian First Team Selection Test, 1993 (34th IMO)

4

Show that for any function $f : \mathcal{P}(\{1, 2, \ldots, n\}) \to \{1, 2, \ldots, n\}$ there exist two subsets, A and B, of the set $\{1, 2, \ldots, n\}$, such that $A \neq B$ and $f(A) = f(B) = \max\{i \mid i \in A \cap B\}$.

Solution

The problem, as stated, is clearly incorrect since for $\max\{i : i \in A \cap B\}$ to make sense, we must have $A \cap B \neq \emptyset$. For $n = 1$ clearly there are no subsets A and B with $A \neq B$ and $A \cap B \neq \emptyset$. A counterexample when $n = 2$ is provided by setting $f(\emptyset) = f(\{1\}) = f(\{2\}) = 1$ and $f(\{1, 2\}) = 2$. This counterexample stands if max is changed to min. The conclusion is still incorrect if $A \cap B$ is changed to $A \cup B$. A counterexample would be $f(\emptyset) = 2$ and $f(\{1\}) = f(\{2\}) = f\{(1, 2)\} = 1$.

Czechoslovakia Mathematical Olympiad, 1993

Find all functions $f : \mathbb{Z} \to \mathbb{Z}$ such that $f(-1) = f(1)$ and
$$f(x) + f(y) = f(x + 2xy) + f(y - 2xy)$$
for all integers x, y.

Solution

$$f(1) + f(y) = f(1 + 2y) + f(-y) \quad (1)$$
$$f(y) + f(-1) = f(-y) + f(-1 + 2y) \quad (2)$$

$f(1) = f(-1)$ and $f(y) = f(y)$, so equating (1) and (2)
$$f(1 + 2y) + f(-y) = f(-y) + f(-1 + 2y).$$
Hence $f(2y + 1) = f(2y - 1)$ for all integers y. $\quad (\star)$

Also $f(y) + f(1) = f(3y) + f(1 - 2y)$ and since $f(1) = f(1 - 2y)$ (making use of $(\star)$) $f(y) = f(3y)$.

Also, equating this with equation (1)
$$f(3y) + f(1 - 2y) = f(1 + 2y) + f(-y),$$
and since by $(\star)$ $f(1 - 2y) = f(1 + 2y)$
$$f(3y) = f(-y).$$
Thus $f(y) = f(3y) = f(-y)$ and $f(y) = f(-y)$ for all y.

If y is odd, then $f(y) = f(y - 2xy)$ so $f(x) = f(x + 2xy)$.

Thus $f(2a) = f(2a(1 + 2y))$.

Thus an odd multiple of a number and that number give the same value.

Therefore if $n = 2^k a$, a odd, then $f(n) = f(2^k)$. Therefore all functions must be such that $f(2^k a) = f(2^k)$, $k \geq 0$ where a is odd and $f(2^k)$ for each k can take any value as can $f(0)$.

These functions always satisfy the conditions since $f(-1) = f(1)$ and
$$f(x + 2xy) + f(y - 2xy) = f(x(1 + 2y)) + f(y(1 - 2x))$$
$$= f(x) + f(y).$$

Mathematical Contest Baltic Way, 1992

Let $a = \sqrt[1992]{1992}$. Which number is greater:

$$\left.\begin{array}{r} a^{\cdot^{\cdot^{\cdot^{a}}}} \\ a \\ a \\ a \end{array}\right\} 1992 \quad \text{or} \quad 1992?$$

Solution

Let $f(x) = a^x$. Clearly $p > q$ if and only if $f(p) > f(q)$; that is, f is strictly increasing. Similarly if $f^n(x)$ is defined by $f^1(x) = f(x)$ and $f^{n+1}(x) = f(f^n(x))$, so $f^n(x) = \underbrace{f\,f\,\ldots\,f(x)}_{n}$ then $p > q$ if and only if $f^n(p) > f^n(q)$ for all n.

Now $1992 > a$.

Thus $f^{1992}(1992) > f^{1992}(a)$. But $f(1992) = 1992$, so $f^{1992}(1992) = 1992$. Now $f^{1992}(a)$ is the expression in question so 1992 is the larger.

Editor's Note. Selby points out that if $x_k = f^k(a)$ then $\lim_{k \to \infty} x_k = L$ where $a = L^{1/L}$ or $L = 1992$.

Mathematical Contest Baltic Way, 1992

A polynomial $f(x) = x^3 + ax^2 + bx + c$ is such that $b < 0$ and $ab = 9c$. Prove that the polynomial has three different real roots.

Solution

Suppose $a = 0$. Then $c = 0$ and $f(x) = x^3 + bx$. Therefore there are three roots: $x = 0$, $x = \sqrt{-b}$, $x = -\sqrt{-b}$. If $a > 0$, then $c = \frac{ab}{9} < 0$, since $b < 0$. Now $f(0) = c < 0$ and since $\lim_{x \to \infty} f(x) = \infty$, there is some r_1 in $(0, \infty)$ such that $f(r_1) = 0$.

Also $f(-a) = -a^3 + a^3 - ab + c = \frac{-8ab}{9} > 0$.

Since $f(0) < 0$ and $f(-a) > 0$, there is a root r_2 in the interval $(-a, 0)$. Finally, since $f(x) \to -\infty$ as $x \to -\infty$ and $f(-a) > 0$ there must be a root r_3 in $(-\infty, -a)$.

If $a < 0$, $c = \frac{ab}{9} > 0$. Hence $f(0) = c > 0$ while $f(x) \to -\infty$ as $x \to -\infty$. Therefore we have a root t_1 in $(-\infty, 0)$.

$f(-a) = -a^3 + a^3 - ab + c = -\frac{8}{9}ab < 0$. Hence there must be a root t_2 in $(0, -a)$. Finally $f(x) \to \infty$ as $x \to \infty$. Since $f(-a) < 0$ there need be a root t_3 in $(-a, \infty)$.

In all cases we have three distinct roots.

Mathematical Contest Baltic Way, 1992

Find all fourth degree polynomials $p(x)$ such that the following four conditions are satisfied:

(i) $p(x) = p(-x)$, for all x,

(ii) $p(x) \geq 0$, for all x,

(iii) $p(0) = 1$,

(iv) $p(x)$ has exactly two local minimum points x_1 and x_2 such that $|x_1 - x_2| = 2$.

Solution

Condition (i) implies $p(x)$ is even.

$$p(x) = ax^4 + bx^3 + cx^2 + dx + e,$$

$$p(-x) = ax^4 - bx^3 + cx^2 - dx + e.$$

Hence we have $2bx^3 + 2dx = 0$ for all x. Thus $b = d = 0$.

Also $p(0) = 1$. Therefore $p(x)$ has the form

$$p(x) = ax^4 + cx^2 + 1.$$

Now $p'(x) = 4ax^3 + 2cx$. The critical points are $x = 0$, $x^2 = \frac{-c}{2a}$. Hence we must have $\frac{-c}{2a} > 0$.

$p''(x) = 12ax^2 + 2c$. Since we want exactly two local minima, we must have at $x^2 = \frac{-c}{2a}$, $p'(x) = 12a\left(\frac{-c}{2a}\right) + 2c = -4c > 0$. Therefore $c < 0$ and $a > 0$.

Further we want $p(x) \geq 0$. Since $p(x) \to \infty$ as $|x| \to \infty$ then since $p(0) = 1$, we need at $x^2 = \frac{-c}{2a}$, $p(x) = a\frac{c^2}{4a^2} + c\left(\frac{-c}{2a}\right) + 1 \geq 0$. So $\frac{-c^2}{4a} + 1 \geq 0$ or $4a \geq c^2$. Also $x_1 = \sqrt{\frac{-c}{2a}}$, $x_2 = -\sqrt{\frac{-c}{2a}}$, $|x_1 - x_2| = 2 \Rightarrow \left|\sqrt{\frac{-c}{2a}}\right| = 1$ or $-c = 2a$. Since $4a \geq c^2$, we have $4a \geq 4a^2$ or $a(1-a) \geq 0$.

Therefore since $a > 0$, $1 - a \geq 0$ and we have $0 < a \leq 1$, $c = -2a$.

The polynomials are of the form $ax^4 - 2ax^2 + 1$ where $0 < a \leq 1$.

Mathematical Contest Baltic Way, 1992

Let $\mathbb{Q}^+$ denote the set of positive rational numbers. Show that there exists one and only one function $f : \mathbb{Q}^+ \to \mathbb{Q}^+$ satisfying the following conditions:

(i) If $0 < q < \frac{1}{2}$ then $f(q) = 1 + f\left(\frac{q}{1-2q}\right)$.
(ii) If $1 < q \le 2$ then $f(q) = 1 + f(q-1)$.
(iii) $f(q) \cdot f(\frac{1}{q}) = 1$ for all $q \in \mathbb{Q}^+$.

Solution

By a change of variable $\tilde{q} = \frac{1}{1-2q}$, we have from (i),

$$f\left(\frac{\tilde{q}}{1+2\tilde{q}}\right) = 1 + f(\tilde{q}), \quad (0 < \tilde{q} < \infty), \quad \text{or} \quad f\left(\frac{1}{\frac{1}{\tilde{q}}+2}\right) = 1 + f\left(\frac{1}{\frac{1}{\tilde{q}}}\right).$$

Calling $t = \frac{1}{\tilde{q}}$ and using (iii) we have

$$\frac{1}{f(t+2)} = 1 + \frac{1}{f(t)}, \quad 0 < t < \infty, \quad t \in \mathbb{Q}^+. \tag{1}$$

Then

$$\frac{1}{f(t+4)} = \frac{1}{f(t+2+2)} = 1 + \frac{1}{f(t+2)} = 1 + 1 + \frac{1}{f(t)}$$
$$= 2 + \frac{1}{f(t)}.$$

Hence, we can evaluate $f(t+2k)$, $k \ge 0$, k an integer, if we know $f(t)$.

Observe that condition (ii) can be rewritten as $f(1+t) = 1 + f(t)$, $t \in \mathbb{Q}^+$, $0 < t \le 1$.

We can now evaluate $f(2k+1+q)$ as follows: Since

$$\frac{1}{f(2+q)} = 1 + \frac{1}{f(q)}, \quad \text{we have} \quad \frac{1}{f(2+1+q)} = 1 + \frac{1}{f(1+q)}.$$

If $0 < q \le 1$, then $\frac{1}{f(3+q)} = 1 + \frac{1}{1+f(q)}$. Hence $f(3+q)$, $0 < q \le 1$ can be evaluated if $f(q)$ is known. Once $f(3+q)$ is known, we obtain

$$\frac{1}{f(5+q)} = \frac{1}{f(2+3+q)} = 1 + \frac{1}{f(3+q)},$$

and

$$\frac{1}{f(2k+1+q)} + 1 + \frac{1}{f(2k-1+q)}, \quad 1 \ge q > 0.$$

Therefore, we can now evaluate

$$f(2k+q), \, f(2k+1+q) \quad 0 < q \le 1, \tag{2}$$

for all $k \ge 0$, k an integer, if we know $f(q)$.

Furthermore, we can evaluate $f(n)$, $n \ge 1$.

First $f(1) = 1$ since putting $q = 1$ in (iii) gives $(f(1))^2 = 1$. Now $f(2) = 1 + f(1) = 2$ from (ii). We follow recursively, $f(3)$:

$$\frac{1}{f(3)} = 1 + \frac{1}{f(1)} = 2$$

and

$$\frac{1}{f(2k+1)} = 1 + \frac{1}{f(2k-1)}.$$

Similarly

$$\frac{1}{f(2k+2)} = 1 + \frac{1}{f(2k)} \text{ and } f(2) = 2.$$

Thus any such function is uniquely defined on the integers.

Finally, we can evaluate the function at any q from the values on the positive integers. Let $q = \frac{a}{b}$, where $(a, b) = 1$.

Write $a = bq_1 + r_1$ where q_1 is a non-negative integer, and $0 \le r_1 < b$ is an integer. If $r_1 = 0$, $f(q) = f(q_1)$ which is determined.

If $a \le r_1 < b$, we apply $f(\frac{a}{b}) = f(q_1 + \frac{r_1}{b})$. This is determined if the value of $f(\frac{r_1}{b})$ is known using (2). Now $0 < \frac{r_1}{b} < 1$. We now compute $f(\frac{b}{r_1})$. $b = r_1 q_2 + r_2$, $r_2 < r_1$. Continuing, since $0 \le r_{k+1} < r_k$, $r_j = 0$ for some j, and we will have an expression for which f is evaluated at an integer. Hence f exists and is uniquely determined.

Mathematical Contest Baltic Way, 1992

Let $\mathbb{N}$ denote the set of positive integers. Let $\varphi : \mathbb{N} \to \mathbb{N}$ be a bijective function and assume that there exists a finite limit

$$\lim_{n \to \infty} \frac{\varphi(n)}{n} = L.$$

What are the possible values of L?

Solution

We claim L must be 1.

Consider $\max\{\varphi(1), \ldots, \varphi(n)\} = j_n$. We note that $j_n \geq n$, since φ is one-to-one. Let $i_n \in \{1, 2, \ldots, n\}$ be such that $\varphi(i_n) = j_n$. Then

$$\frac{\varphi(i_n)}{i_n} \geq 1.$$

Since

$$\lim_{n \to \infty} \frac{\varphi(n)}{n} = L, \quad \lim_{n \to \infty} \frac{\varphi(i_n)}{i_n} = L \geq 1. \qquad (1)$$

Now consider $S_n = \{n \in N : \varphi(n) \leq n\}$. S_n must be infinite. First $S_n \neq \emptyset$ for if $S_n = \emptyset$ then $\varphi(k) > k$ for all k and there is no k_0 with $\varphi(k_0) = 1$.

Suppose S_n is finite, with k the largest value in the set. Then $\varphi(n) > n$ for $n \geq k+1$. Consider $\{1, 2, \ldots, k+1\}$. Since $\varphi(n) > k+1$ for $n \geq k+1$, the only integers which can be pre-images of $\{1, 2, \ldots, k+1\}$ are $\{1, 2, \ldots, k\}$. This is not possible, since φ is one-to-one and onto.

Therefore $S_n = \{n \in N : \varphi(n) \leq n\}$ is infinite. Choose a sequence, $n_k \in S_n$ with $n_k \to \infty$. We now have $\lim\limits_{k \to \infty} \frac{\varphi(n_k)}{n_k} = L$. However

$$\frac{\varphi(n_k)}{n_k} \leq 1.$$

Thus

$$L \leq 1. \qquad (2)$$

From (1) and (2), $L = 1$.

3rd Mathematical Olympiad of the Republic of China, 1994 (First day)

Let a, b, c be positive real numbers, α be a real number. Suppose that
$$f(\alpha) = abc(a^\alpha + b^\alpha + c^\alpha)$$
$$g(\alpha) = a^{\alpha+2}(b+c-a) + b^{\alpha+2}(a-b+c) + c^{\alpha+2}(a+b-c)$$
Determine the magnitude between $f(\alpha)$ and $g(\alpha)$.

Solution

$$bca^{\alpha+1} + acb^{\alpha+1} + abc^{\alpha+1} - a^{\alpha+2}(b+c-a)$$
$$-b^{\alpha+2}(a-b+c) - c^{\alpha+2}(a+b-c)$$
$$= a^{\alpha+1}(bc - a(b+c) + a^2) + b^{\alpha+1}(ac - b(a+c) + b^2)$$
$$+ c^{\alpha+1}(ab - c(a+b) + c^2)$$
$$= a^{\alpha+1}(a-b)(a-c) + b^{\alpha+1}(b-a)(b-c) + c^{\alpha+1}(c-a)(c-b)$$
$$\geq 0$$

which is an inequality of Schur.

Israel Mathematical Olympiad, 1994

p and q are positive integers. f is a function defined for positive numbers and attains only positive values, such that $f(xf(y)) = x^p y^q$. Prove that $q = p^2$.

Solution

For $x = \dfrac{1}{f(y)}$, we get $f(y) = \dfrac{y^{q/p}}{(f(1))^{1/p}}$.

For $y = 1$, we get $f(1) = 1$, so $f(y) = y^{q/p}$. Hence $f(x \cdot y^{q/p}) = x^p \cdot y^q$.
For $y = z^{p/q}$ we get $f(x \cdot z) = x^p z^p$ or $f(x) = x^p$.

Thus $\dfrac{q}{p} = p$, whence $q = p^2$.

Israel Mathematical Olympiad, 1994

Find all real coefficients polynomials $p(x)$ satisfying

$$(x-1)^2 p(x) = (x-3)^2 p(x+2)$$

for all x.

Solution

We consider polynomials $p(x)$ with coefficients in a field $\mathbb{F}$ of *arbitrary* characteristic and find as follows:

(i) If $\text{char}(\mathbb{F}) = 0$, (in particular, if $\mathbb{F} = \mathbb{R}$), then $p(x) = a(x-3)^2$, where a is any scalar (possibly 0) in $\mathbb{F}$;

(ii) If $\text{char}(\mathbb{F}) = 2$, then *every* $p(x)$ satisfies the equation (clear);

(iii) If $\text{char}(\mathbb{F}) =$ an odd prime, l, then there are infinitely many solutions, including all $p(x) = a(x-3)^2(x^{l^\nu} - x + c)$ with $a, c \in \mathbb{F}$, and $\nu = 0, 1, 2, \ldots$. (Note that $p(x)$ has the form $a(x-3)^2$ if $\nu = 0$.)

To prove this, observe that if $\text{char}(\mathbb{F}) \neq 2$, then $x-1$ and $x-3$ are coprime, whence $p(x) = (x-3)^2 q(x)$ in $\mathbb{F}[x]$.

Thus our equation becomes

$$(x-1)^2 (x-3)^2 q(x) = (x-3)^2 (x-1)^2 q(x+2) \quad (*)$$

whence $q(x) = q(x+2)$, as polynomials; that is, elements of $\mathbb{F}[x]$.

Now if $\text{char}(\mathbb{F}) = 0$, then $(*)$ has only constant solutions.

(The most elementary proof of this: without loss of generality, $q(x) = x^n + ax^{n-1} + \cdots$. Then $q(x+2) - q(x) = 2nx^{n-1} + \cdots$, and this is non-zero if $n \geq 1$. Another proof: $(*)$ implies that $q(x)$ is periodic, which forces equations $q(x) = c$ to have infinitely many roots x, a contradiction).

This establishes the assertion (i).

Re: assertion (iii). Let $\text{char}(\mathbb{F}) = l$ and $q(x) = x^{l^\nu} - x + c$.

Then for $x = 0, 1, \ldots, l-1$, (that is for each element of the prime field), we have $q(x) = c$ and so $q(x) = q(x+1) = q(x+2) = \ldots$, yielding polynomials of degree greater than or equal to l which satisfy $(*)$. This establishes the assertion (iii).

Swedish Mathematical Olympiad, 1993

Let a and b be real numbers and let $f(x) = (ax+b)^{-1}$. For which a and b are there three distinct real numbers x_1, x_2, x_3 such that $f(x_1) = x_2$, $f(x_2) = x_3$ and $f(x_3) = x_1$?

Solution

Consider the functions of the form

$$g(x) = \frac{\alpha x + \beta}{\gamma x + \delta}.$$

Lemma. $g(x)$ has at least 3 distinct fixed points if and only if $\gamma = \beta = 0, \alpha = \delta \neq 0$.

Proof. If $\gamma = \beta = 0, \alpha = \delta \neq 0$, $g(x) = x$ and it clearly has at least 3 distinct points x_1, x_2, x_3 such that $g(x_i) = x_i$, $i = 1, 2, 3$. Conversely consider the equation for a fixed point x, $g(x) = x$. This implies $\gamma x^2 + \delta x = \alpha x + \beta$ or $\gamma x^2 + (\delta - \alpha)x - \beta = 0$. Suppose this has three distinct roots. Then the quadratic must be identically 0, or $\gamma = \beta = 0$ and $\alpha = \delta$.

Now, if $f(x) = \frac{1}{ax+b}$, then

$$f \circ f(x) = \frac{ax+b}{abx + b^2 + a} \quad \text{and} \quad f \circ f \circ f(x) = \frac{abx + a + b^2}{a(a+b^2)x + ab + b(b^2 + a)}.$$

The problem implies $f \circ f \circ f$ has three distinct real fixed points x_1, x_2, x_3. By the above lemma, this is true if and only if

$$a + b^2 = a(a + b^2) = 0 \quad \text{and} \quad ab + b(a + b^2) = ab \neq 0.$$

This is true if and only if $a = -b^2$ and $ab \neq 0$.

Irish Mathematical Olympiad, 1994

Determine with proof all real polynomials $f(x)$ satisfying the equation
$$f(x^2) = f(x)f(x-1).$$

Solution

We will prove that $f(x) = 0$ or $f(x) = (x^2 + x + 1)^k$, $k = 0, 1, 2, \ldots$. Consider any (possibly complex) root p of $f(x)$. Then
$$f(p^2) = f(p) \cdot f(p-1) = 0 \cdot f(p-1) = 0$$
and
$$f((p+1)^2) = f(p+1) \cdot f(p) = f(p+1) \cdot 0 = 0.$$
So p^2, $(p+1)^2$ are also roots of $f(x)$. Thus p^{2^n} and $(p+1)^{2^n}$ are roots of $f(x)$, $n = 0, 1, 2, \ldots$. If $|p| \neq 1$ or $|p| \neq |p+1|$ then we get an infinite number of roots, so $f(x)$ is a constant polynomial, and having a root p, $f(x) \equiv 0$.

If $|p| \neq 1$ or $|p+1| \neq 1$; that is, if $|p| = 1 = |p+1|$ then $p \cdot \overline{p} = 1$ and $p \cdot \overline{p} = (p+1)(\overline{p}+1)$, so $p + \overline{p} = -1$ and $\overline{p} = -(p+1)$, and now $p(-p-1) = 1$. Therefore $p^2 + p + 1 = 0$. It follows that $f(x) = (x^2+x+1)^k$, for some $k \geq 1$.

On the other hand if $f(x)$ has no roots, then $f(x) = c \neq 0$, is a non-zero constant. Then $f(x^2) = f(x) \cdot f(x-1)$ gives $c = c \cdot c$, and $c \neq 0$ gives $c = 1$. Thus $f(x) = (x^2 + x + 1)^0$. In any case $f(x) = 0$ or $f(x) = (x^2 + x + 1)^k$ for some $k = 0, 1, 2, \ldots$.

37th International Mathematical Olympiad, 1996 (Shortlist)

Let f be a function from the set of real numbers $\mathbb{R}$ into itself such that for all $x \in \mathbb{R}$, we have $|f(x)| \leq 1$ and

$$f\left(x + \frac{13}{42}\right) + f(x) = f\left(x + \frac{1}{6}\right) + f\left(x + \frac{1}{7}\right).$$

Prove that f is a periodic function (that is, there exists a non-zero real number c such that $f(x + c) = f(x)$ for all $x \in \mathbb{R}$).

Solution

We will prove that f is 1-periodic. We have

$$f\left(x + \frac{k}{6} + \frac{l+1}{7}\right) + f\left(x + \frac{k+1}{6} + \frac{l}{7}\right)$$
$$= f\left(x + \frac{k}{6} + \frac{l}{7}\right) + f\left(x + \frac{k+1}{6} + \frac{l+1}{7}\right).$$

If k runs through $1, 2, \ldots, m-1$, where $m \in \mathbb{N}$, and adding these equations we obtain

$$f\left(x + \frac{l+1}{7}\right) + f\left(x + \frac{m}{6} + \frac{l}{7}\right) = f\left(x + \frac{m}{6}\right) + f\left(x + \frac{m}{6} + \frac{l+1}{7}\right).$$

Similarly when l runs $1, 2, \ldots, n-1$, ($n \in \mathbb{N}$) and adding these equations, we obtain

$$f\left(x + \frac{n}{7}\right) + f\left(x + \frac{m}{6}\right) = f(x) + f\left(x + \frac{m}{6} + \frac{n}{7}\right).$$

We choose $n = 7$ and $m = 6$, and find

$$2f(x+1) = f(x) + f(x+2).$$

This means that the sequence $f(x+n)$ is an arithmetic sequence with common difference $f(x+1) - f(x)$. But, since f is bounded, we must have $f(x+1) - f(x) = 0$. Hence 1 is a period of f. Finally, 1 is the "best" period because the function

$$f(x) = \frac{\{6x\} + \{7x\}}{2}$$

satisfies all the hypotheses of the problem. (Here, $\{x\}$ denotes the fractional part of x.)

Croatian National Mathematics Competition, 1994 (4th Class)

For a complex number z let $w = f(z) = \dfrac{2}{3-z}$.

(a) Determine the set $\{w : z = 2 + iy, y \in \mathbb{R}\}$ in the complex plane.

(b) Show that the function w can be written in the form

$$\frac{w-1}{w-2} = \lambda \frac{z-1}{z-2}.$$

(c) Let $z_0 = \frac{1}{2}$ and the sequence (z_n) be defined recursively by

$$z_n = \frac{2}{3 - z_{n-1}}, \quad n \geq 1.$$

Using the property (b) calculate the limit of the sequence (z_n).

Solution

(a) Let $w = u + iv$, where $u, v \in \mathbb{R}$. Then from $w = \dfrac{2}{1-iy} = \dfrac{2(1+iy)}{1+y^2}$ we get $u = \dfrac{2}{1+y^2}$ and $v = \dfrac{2y}{1+y^2}$. Eliminating y, we get $u^2 + v^2 = \dfrac{4}{1+y^2} = 2u$, or $(u-1)^2 + v^2 = 1$. Hence $\{w : z = 2 - iy, y \in \mathbb{R}\}$ is the set of all points on the circle centred at $(1,0)$ with radius 1.

(b) Straightforward computations show that

$$\frac{w-1}{w-2} = \left(\frac{z-1}{3-z}\right) \div \left(\frac{2z-4}{3-z}\right) = \frac{1}{2}\left(\frac{z-1}{z-2}\right);$$

that is, $\lambda = \frac{1}{2}$.

(c) Using the equation in (b), we find, by iteration, that

$$\frac{z_n - 1}{z_n - 2} = \frac{1}{2}\left(\frac{z_{n-1}-1}{z_{n-1}-2}\right) = \left(\frac{1}{2}\right)^n \left(\frac{z_0 - 1}{z_0 - 2}\right) = \frac{1}{3}\left(\frac{1}{2}\right)^n.$$

Since $(\frac{1}{2})^n \to 0$ as $n \to \infty$, we find $\lim_{n \to \infty} z_n = 1$.

Croatian National Mathematics Competition, 1994 (4th Class)

3

Determine all polynomials $P(x)$ with real coefficients such that for some $n \in \mathbb{N}$ we have $xP(x-n) = (x-1)P(x)$, for all $x \in \mathbb{R}$.

Solution

The only such polynomials are $P(x) = cx$, where c is an arbitrary constant. Clearly $P(x) \equiv 0$ satisfies the given equation. So assume $P(x) \not\equiv 0$. Setting $x = 0$ in $xP(x-n) = (x-1)P(x)$, we get $P(0) = 0$ and thus $(n-1)P(n) = 0$. If $n \neq 1$, then $P(n) = 0$ and a straightforward induction shows that $P(kn) = 0$ for all $k \in \mathbb{N}$, which is impossible since $P(x) \not\equiv 0$. Hence $n = 1$ and we have $xP(x-1) = (x-1)P(x)$ which implies $P(2) = 2P(1)$. Suppose that $P(m) = mP(1)$ for some $m \geq 2$. Then from $(m+1)P(m) = mP(m+1)$, we get $P(m+1) = (m+1)P(1)$. Hence $P(m) = mP(1) = mc$ for all $m \in N$, where $c = P(1)$. Let $Q(x) = P(x) - cx$. Then $Q(m) = P(m) - cm = 0$ for all $m \in N$ and so $Q(x) \equiv 0$ from which $P(x) \equiv cx$. Noting that $c = 0$ if and only if $P(x) \equiv 0$, the conclusion follows.

17th Austrian-Polish Mathematics Competition, 1994

The function $f : \mathbb{R} \to \mathbb{R}$ satisfies for all $x \in \mathbb{R}$ the conditions

$$f(x+19) \leq f(x) + 19 \quad \text{and} \quad f(x+94) \geq f(x) + 94.$$

Show that $f(x+1) = f(x) + 1$ for all $x \in \mathbb{R}$.

Solution

Let x be an arbitrary real number. Applying the given conditions to $x - 19$ and $x - 94$ respectively, we obtain

$$f(x-19) \geq f(x) - 19 \quad \text{and} \quad f(x-94) \leq f(x) - 94.$$

Now an easy induction shows that for all $n \in \mathbb{N}$,

$$f(x+19n) \leq f(x) + 19n, \quad f(x+94n) \geq f(x) + 94n,$$
$$f(x-19n) \geq f(x) - 19n, \quad \text{and} \quad f(x-94n) \leq f(x) - 94n.$$

Since $1 = 5 \times 19 - 94$ and $1 = 18 \times 94 - 89 \times 19$, we get:

$$\begin{aligned} f(x+1) = f(x + 5 \times 19 - 94) &\leq f(x + 5 \times 19) - 94 \\ &\leq f(x) + 5 \times 19 - 94 \\ &= f(x) + 1, \end{aligned}$$

and

$$\begin{aligned} f(x+1) = f(x + 18 \times 94 - 89 \times 19) &\geq f(x + 18 \times 94) - 89 \times 19 \\ &\geq f(x) + 18 \times 94 - 89 \times 19 \\ &= f(x) + 1, \end{aligned}$$

so that $f(x+1) = f(x) + 1$, as required.

37th International Mathematical Olympiad, 1996 (Shortlist)

Let n be an even positive integer. Prove that there exists a positive integer k such that

$$k = f(x)(x+1)^n + g(x)(x^n + 1)$$

for some polynomials $f(x)$, $g(x)$ having integer coefficients. If k_0 denotes the least such k, determine k_0 as a function of n.

Solution

The statement of the problem is equivalent to: prove that for any positive integer n, there exist polynomials $f(x)$, $g(x)$ having integer coefficients such that
$$f(x)(x+1)^{2n} + g(x)(x^{2n} + 1) = 2,$$
or equivalently to find $f(x)$ and $g(x)$ such that
$$f(x)x^{2n} + g(x)((x-1)^{2n} + 1) = 2,$$
or equivalently, to find $g(x)$ such that $g(x)((x-1)^{2n} + 1) - 2$ is divisible by x^{2n}.

But, this is not difficult to prove. If
$$g(x) = a_0 + a_1 x + \cdots + a_{2n-1} x^{2n-1},$$
then we have only to choose a_i, $0 \leq i \leq 2n - 1$ as follows:

$$a_0 = 1 \quad \text{and} \quad 2a_k - \binom{2n}{1} a_{k-1} + \binom{2n}{2} a_{k-2} + \cdots + (-1)^k \binom{2n}{k} = 0$$

for $1 \leq k \leq 2n - 1$.

37th International Mathematical Olympiad, 1996 (Shortlist)

Let the sequence $a(n)$, $n = 1, 2, 3, \ldots$, be generated as follows: $a(1) = 0$, and for $n > 1$,

$$a(n) = a(\lfloor n/2 \rfloor) + (-1)^{n(n+1)/2}.$$

(Here $\lfloor t \rfloor$ = the greatest integer $\leq t$.)

(a) Determine the maximum and minimum value of $a(n)$ over $n \leq 1996$, and find all $n \leq 1996$ for which these extreme values are attained.

(b) How many terms $a(n)$, $n \leq 1996$, are equal to 0?

Solution

A simple induction shows that $a(n) = b(n) - c(n)$ where $b(n)$ (resp. $c(n)$) denote the total number of couplings 00, 11 (resp. 01, 10) in the binary representation of n.

(To see this, it suffices to observe that the binary representation of $\lfloor \frac{n}{2} \rfloor$ is obtained from that of n by deleting the last digit.)

(a) The maximum value of $a(n)$ is the largest $n \leq 1996$ for which $c(n) = 0$. It is attained for 1111111111 or 1023 where $b(n) = 9$ and hence $a(n) = 9$. The minimum is the largest $n \leq 1996$ for which $b(n) = 0$. It is attained for 10101010101 or 1365 where $a(n) = -10$.

(b) $a(n) = 0$ if and only if, in the binary representation of n, the number of the same two consecutive digits is equal to the number of different two consecutive digits. Noting that the first digit has to be 1 and that such representation of n can be formed in $\binom{m}{m/2}$ ways for even m, we deduce that there are

$$\binom{0}{0} + \binom{2}{1} + \binom{4}{2} + \binom{6}{3} + \binom{8}{4} + \binom{10}{5} = 351$$

positive integers $n < 2^{11} = 2048$ with $a(n) = 0$. Finally, since 2002, 2004, 2006, 2010, 2026 are such that $a(n) = 0$ but exceed 1996, there are only 346 numbers ≤ 1996 with $a(n) = 0$.

Iranian National Mathematical Olympiad, 1994 (Second Round)

$f(x)$ and $g(x)$ are polynomials with real coefficients such that for infinitely many rational values x, $\frac{f(x)}{g(x)}$ is rational. Prove that $\frac{f(x)}{g(x)}$ can be written as the ratio of two polynomials with rational coefficients.

Solution

With little change, a solution of this problem is given in [1]. Without loss of generality we can assume that f and g are relatively prime polynomials and let r denote the sum of the degrees of f and g. Also let $R(x) = f(x)/g(x)$, assuming that the degree of f is $\geq$ than the degree of g. If not, we consider $R^{-1}(x)$. For $r = 0$, the result is obvious. Now let a be one of the rational numbers such that $g(a) \neq 0$ and $R(a)$ is rational. Now define $f_1(x)$ by

$$f_1(x)/g(x) = \{R(x) - f(a)/g(a)\}/(x-a).$$

Then $f_1(x)/g(x)$ is also rational for all the rational values that $f(x)/g(x)$ is rational except $x = a$. Since

$$f_1(x) = [g(a)f(x) - f(a)g(x)]/g(a)(x-a),$$

the degree of $f_1(x)$ is less than that of $f(x)$. Thus the sum of the degrees of $f_1(x)$ and $g(x)$ is less than that of $f(x)$ and $g(x)$. The desired result now follows by mathematical induction.

Balkan Mathematical Olympiad, 1994

[*Greece*] Show that the polynomial

$$x^4 - 1994x^3 + (1993 + m)x^2 - 11x + m, \quad m \in \mathbb{Z}$$

has at most one integral root.

Solution

Consider

$$x^4 - 1994x^3 + (1993 + m)x^2 - 11x + m. \qquad (*)$$

Suppose the given polynomial has two integral roots. Then neither can be odd for otherwise

$$(x^4 + 1993x^2) - (1994x^3) + m(x^2 + 1) - 11x$$

will be odd (as each of the terms in brackets is even) and hence non-zero.

Suppose $x_1 = 2^{r_1}a$, $r_1 > 0$ and a odd is a solution.

Considering the polynomial $\pmod{2^{2r_1}}$, we then have that $m \equiv 11x \pmod{2^{2r_1}}$. Hence $m \equiv 2^{r_1}(11a) \pmod{2^{2r_1}}$, and since a is odd, m must be of the form $2^{r_1}l_1$, l_1 odd.

If $x_2 = 2^{r_2}b$, b odd, is also a solution, then $m = 2^{r_2}l_2$, l_2 odd, so we must have $r_1 = r_2$.

Thus, if there are two integral roots, both must be of the form $2^r k$, $r > 1$, k odd. The product of the two roots must be a multiple of 2^{2r}. The quadratic which has these two roots as zeros is $x^2 + px + 2^{2r}q$, where p, q are integers.

Now the given polynomial (*) can be factorized into two quadratics:

$$(x^2 + px + 2^{2r}q)(x^2 + sx + t).$$

If s were not integral, then the coefficient of x^3 would not be integral in the quartic, and if t were not integral, the coefficient of x^2 would not be integral in the quartic. Thus s and t must be integers and $m = 2^{2r}qt$. But the highest power of 2 dividing m is 2^r so $r \geq 2r$ giving $r = 0$, a contradiction. Hence the given quartic cannot have more than one integral root.

38th Mathematics Competition of the Republic of Slovenia (4th Grade)

1. Prove that there does not exist a function $f : \mathbb{Z} \to \mathbb{Z}$, for which $f(f(x)) = x + 1$ for every $x \in \mathbb{Z}$.

Solution

Suppose that there is such a function. Then $f(f(f(x))) = f(x) + 1$. Since $f(f(x)) = x + 1$, we get $f(x + 1) = f(x) + 1$.

By induction $f(x + n) = f(x) + n$ for every $n \in \mathbb{N}$. Also $f(x) = f(x - n + n) = f(x - n) + n$ so

$$f(x - n) = f(x) - n \quad \text{for every} \quad n \in \mathbb{N}.$$

Finally $f(x + y) = f(x) + y$ for $x, y \in \mathbb{Z}$.

For $x = 0$, $f(y) = f(0) + y$. For $y = f(0)$, $f(f(0)) = f(0) + f(0)$.

But $f(f(0)) = 1$; thus $2f(0) = 1$, a contradiction.

Israel Mathematical Olympiad, 1995

Let n be a given positive integer. A_n is the set of all points in the plane, whose x and y coordinates are positive integers between 0 and n. A point (i,j) is called "internal" if $0 < i,j < n$. A real function f, defined on A_n, is called a "good function" if it has the following property: for every internal point x, the value of $f(x)$ is the mean of its values on the four neighbouring points (the neighbouring points of x are the four points whose distance from x equals 1). f and g are two given good functions and $f(a) = g(a)$ for every point a in A_n which is not internal (that is, a boundary point). Prove that $f \equiv g$.

Solution

Let ∂A_n denote the boundary of A_n. By hypothesis $f = g$ on ∂A_n, and our goal is to prove that $f = g$ everywhere. To this end we will prove that $\min(f - g) = 0 = \max(f - g)$.

Assume that $\max(f - g)$ is achieved at a point m_0. If m_0 is an internal point, then the value of $f - g$ is the mean of its value on the four neighbouring points of m_0. But these have values at most equal to $\max(f - g)$. So, each of them achieves the maximum.

However, we conclude that the maximum is achieved at a point of ∂A_n, and then $\max(f - g) = 0$.

Similar arguments give $\min(f - g) = 0$.

Finally, we conclude $f \equiv g$.

Israel Mathematical Olympiad, 1995

α is a given real number. Find all functions $f : (0, \infty) \mapsto (0, \infty)$ such that the equality

$$\alpha x^2 f\left(\frac{1}{x}\right) + f(x) = \frac{x}{x+1}$$

holds for all real $x > 0$.

Solution

We have
$$\alpha x^2 f\left(\frac{1}{x}\right) + f(x) = \frac{x}{x+1}$$
$$\alpha \frac{1}{x^2} f(x) + f\left(\frac{1}{x}\right) = \frac{1}{x+1}$$

(with $\frac{1}{x}$ instead of x) and hence

$$\begin{cases} \alpha x f(\frac{1}{x}) + \frac{1}{x} f(x) = \frac{1}{x+1} \\ \alpha \frac{1}{x} f(x) + x f(\frac{1}{x}) = \frac{x}{x+1} \end{cases}.$$

If $\alpha^2 \neq 1$, then we obtain

$$f(x) = \frac{x(1-\alpha x)}{(x+1)(1-\alpha^2)}$$

and if $\alpha^2 = 1$, there is no solution.

Since $f(x) > 0$, it is easy to see that α must belong to $(-1, 0)$.

45th Latvian Mathematical Olympiad, 1994 (12th Grade)

Does there exist a polynomial $P(x, y)$ in two variables such that
(a) $P(x, y) > 0$ for all x, y?
(b) for each $c > 0$ there exist x and y such that $P(x, y) = c$?

Solution

Yes! Let $P(x, y) = (y^2+1)x^2 + 2xy + 1$. For a fixed y, $P(x, y) = f_y(x)$ a polynomial in x of degree two with discriminant $\Delta = -4$. Hence for all x, y, $P(x, y) > 0$.

For a fixed y, $f_y(\mathbb{R}) = [M(y), +\infty)$ where $M(y) = \frac{1}{y^2+1}$. We have $\lim_{y \to +\infty} M(y) = 0$.

Then, for each $c > 0$ there is $y_0 \in \mathbb{R}$ such that $c \geq M(y_0)$, so there is x_0 with $P(x_0, y_0) = c$.

4th Mathematical Olympiad of the Republic of China (Taiwan), 1995

Let $P(x) = a_0 + a_1 x + \cdots + a_{n-1} x^{n-1} + a_n x^n$ be a polynomial with complex coefficients. Suppose the roots of $P(x)$ are $\alpha_1, \alpha_2, \ldots, \alpha_n$ with $|\alpha_1| > 1$, $|\alpha_2| > 1$, $\ldots$, $|\alpha_j| > 1$, and $|\alpha_{j+1}| \leq 1$, $\ldots$, $|\alpha_n| \leq 1$. Prove:
$$\prod_{i=1}^{j} |\alpha_i| \leq \sqrt{|a_0|^2 + |a_1|^2 + \cdots + |a_n|^2}.$$

Solution

Let
$$\begin{aligned} Q(x) &= a_0 x^m + a_1 x^{m-1} + \cdots + a_m \\ R(x) &= b_0 x^n + b_1 x^{n-1} + \cdots + b_n \\ Q(x) R(x) &= c_0 x^{n+m} + c_1 x^{n+m-1} + \cdots + c_{n+m} \\ \text{and} \quad Q(x) \overline{R}\left(\tfrac{1}{x}\right) &= d_{-m} x^m + \cdots + d_n x^{-n} \end{aligned}$$

with $a_0 = b_0 = 1$.

We claim that
$$\sum_{i=0}^{m+n} c_i^2 = \sum_{k=-m}^{n} d_k^2. \tag{1}$$

Indeed, d_k is the sum of $a_\alpha b_\beta$ with $\alpha - \delta = k$, so that d_k^2 is the sum of $a_\alpha b_\beta a_\gamma b_\delta$ with $\alpha - \delta = \gamma - \beta = k$, and summing over k means we take all $\alpha, \beta, \gamma, \delta$ with $\alpha - \delta = \gamma - \beta$. Hence

$$\begin{aligned} \sum_{i=0}^{m+n} c_i^2 &= \sum_{\alpha + \beta = \gamma + \delta} a_\alpha b_\beta a_\gamma b_\delta \\ &= \sum_{\alpha - \delta = \beta - \gamma} a_\alpha b_\delta a_\gamma b_\beta = \sum_{k=-m}^{n} d_k^2. \end{aligned}$$

Hence, we set
$$\begin{aligned} Q(x) &= (x - \alpha_1)(x - \alpha_2) \ldots (x - \alpha_j), \\ R(x) &= (x - \alpha_{j+1})(x - \alpha_{j+2}) \ldots (x - \alpha_n). \end{aligned}$$

Because of (1), we know that $|a_0|^2 + \cdots + |a_n|^2$ is equal to the sum of the squares of the absolute values of the coefficients of $Q(x)\overline{R}(\tfrac{1}{x})$, and in particular we get $|\alpha_1 \ldots \alpha_j|^2 \leq |a_0|^2 + \cdots + |a_n|^2$.

4th Mathematical Olympiad of the Republic of China (Taiwan), 1995

Given n distinct integers $m_1, m_2, \ldots, m_n$, prove that there exists a polynomial $f(x)$ of degree n and with integral coefficients which satisfies the following conditions:

(1) $f(m_i) = -1$, for all i, $1 \leq i \leq n$.

(2) $f(x)$ cannot be factorized into a product of two non-constant polynomials with integral coefficients.

Solution

It suffices to show that $f(x) = (x - m_1) \ldots (x - m_n) - 1$ satisfies condition (2).

Suppose, for a contradiction, that $f = QR$, with $Q, R \in \mathbb{Z}[x]$, and Q, R, with degree at most $n - 1$.

For all $i \in \{1, \ldots, n\}$, since $f(m_i) = -1 = Q(m_i)R(m_i)$, with $Q(m_i)$, $R(m_i)$ integers, then

$$Q(m_i) = 1 \quad \text{and} \quad R(m_i) = -1, \quad \text{or}$$
$$Q(m_i) = -1 \quad \text{and} \quad R(m_i) = 1.$$

So, $Q + R$ is a polynomial with degree $\leq n-1$ and with n distinct roots, then $Q + R \equiv 0$; that is, $Q \equiv -R$. Thus $f = -Q^2$, which is impossible since the leading coefficient of f is 1, a contradiction. Thus f is irreducible in $\mathbb{Z}[X]$.

18th Austrian-Polish Mathematics Competition, 1995

Let $P(x) = x^4 + x^3 + x^2 + x + 1$. Show that there exist polynomials $Q(y)$ and $R(y)$ of positive degrees, with integer coefficients, such that $Q(y) \cdot R(y) = P(5y^2)$ for all y.

Solution

Since $P(5y^2) = 5^4 y^8 + 5^3 y^6 + 5^2 y^4 + 5y^2 + 1$, we try factors of the form

$$(25y^4 + ay^3 + by^2 + cy + 1)(25y^4 - ay^3 + by^2 - cy + 1).$$

On expanding out, these are factors: $a = 25$, $b = 15$, and $c = 5$.

31st Spanish Mathematical Olympiad, 1994 (First Round)

Let a, b, c be distinct real numbers and $P(x)$ a polynomial with real coefficients. If:

- the remainder on division of $P(x)$ by $x - a$ equals a,
- the remainder on division of $P(x)$ by $x - b$ equals b,
- and the remainder on division of $P(x)$ by $x - c$ equals c;

determine the remainder on division of $P(x)$ by $(x - a)(x - b)(x - c)$.

Solution

As is well known, the remainder on division of $P(x)$ by $x - a$ is $P(a)$. So, the hypotheses imply: $P(a) = a$, $P(b) = b$, $P(c) = c$.

Let $R(x)$ be the remainder on division of $P(x)$ by $(x-a)(x-b)(x-c)$, so that the degree of $R(x)$ is ≤ 2 and $P(x) = (x - a)(x - b)(x - c)Q(x) + R(x)$ for a polynomial $Q(x)$.

We remark that $R(a) = P(a) = a$ and similarly $R(b) = b$ and $R(c) = c$. From this observation, we may conclude through one of the three following ways:

(1) the polynomial $R(x) - x$ has degree ≤ 2 and three distinct zeros a, b, c. Hence $R(x) - x$ is the zero polynomial and $R(x) = x$.

(2) $R(x)$ has the form $ux^2 + vx + w$ where (u, v, w) is the solution of the system

$$\begin{cases} ua^2 + va + w = a \\ ub^2 + vb + w = b \\ uc^2 + vc + w = c. \end{cases} \quad (S)$$

The determinant of (S) is a Vandermonde determinant and is not zero (since a, b, c are distinct), so (S) has a unique solution, which clearly is $u = 0$, $v = 1$, $w = 0$. Thus $R(x) = x$ again.

(3) $R(x)$ is the Lagrange's interpolation polynomial:

$$R(x) = a\frac{(x - b)(x - c)}{(a - b)(a - c)} + b\frac{(x - a)(x - c)}{(b - a)(b - c)} + c\frac{(x - a)(x - b)}{(c - a)(c - b)}.$$

Multiplying out and grouping similar terms, a lengthy but easy calculation provides $R(x) = x$ again.

31st Spanish Mathematical Olympiad, 1994 (First Round)

Show that there exists a polynomial $P(x)$, with integer coefficients, such that $\sin 1°$ is a root of $P(x) = 0$.

Solution

We have
$$e^{i\pi/180} = \cos(\frac{\pi}{180}) + i\sin(\frac{\pi}{180})$$
$$= \cos 1° + i \sin 1°$$
and $(e^{i\pi/180})^{180} = e^{i\pi} = -1$.

Then
$$\left(\cos\frac{\pi}{180} + i\sin\frac{\pi}{180}\right)^{180} = -1.$$

Let $b = \sin\frac{\pi}{180}$, $a = \cos\frac{\pi}{180}$. Then $a^2 = 1 - b^2$ and we have

$$-1 = (a+ib)^{180} = \sum_{k=0}^{180} \binom{180}{k} a^k (ib)^{180-k}.$$

If we take the real parts

$$-1 = \sum_{k=0}^{90} \binom{180}{2k} a^{2k} (-1)^{90-k} b^{180-2k};$$

that is,

$$1 + \sum_{k=0}^{90} (-1)^k \binom{180}{2k} (1-b^2)^{2k} b^{180-2k} = 0,$$

and $\sin 1°$ is a root of a polynomial with integer coefficients.

Vietnamese Mathematical Olympiad, 1996 (Category A)

Determine all functions $f : \mathbb{N}^* \to \mathbb{N}^*$ satisfying:

$$f(n) + f(n+1) = f(n+2)f(n+3) - 1996$$

for every $n \in \mathbb{N}^*$ ($\mathbb{N}^*$ is the set of positive integers).

Solution

From the equation we obtain

$$f(n+1) + f(n+2) = f(n+3)f(n+4) - 1996, \quad (1)$$
$$f(n) + f(n+1) = f(n+2)f(n+3) - 1996. \quad (2)$$

Now, (1) minus (2) yields

$$f(n+2) - f(n) = f(n+3)(f(n+4) - f(n+2)).$$

Hence, $\forall n \in \mathbb{N}^*$,

$$f(3) - f(1) = f(4)f(6)\cdots f(2n+2)\left(f(2n+3) - f(2n+1)\right), \quad (3)$$

$$f(4) - f(2) = f(5)f(7)\cdots f(2n+3)\left(f(2n+4) - f(2n+2)\right). \quad (4)$$

From (3), and if $f(1) > f(3)$, we obtain an infinite decreasing sequence $f(1), f(3), \ldots$ of positive integers, a contradiction. Hence, $f(1) \leq f(3)$.

Case 1: $f(1) = f(3)$.

By (3) we have, $\forall n \in \mathbb{N}^*$,

$$f(2n+3) - f(2n+1) = f(3) - f(1) = 0,$$

and hence,

$$f(2n-1) = f(1). \quad (5)$$

From (4) we get, $\forall n \in \mathbb{N}^*$,

$$f(4) - f(2) = (f(1))^n (f(2n+4) - f(2n+2)). \quad (6)$$

If $f(1) = 1$, by (6), (1) and (5) we obtain $f(4) - f(2) = 1997$ and

$$f(n) = \begin{cases} 1 & \text{if } n \text{ odd,} \\ a + \left(\frac{n}{2} - 1\right) \cdot 1997 & \text{if } n \text{ even,} \end{cases}$$

where $a \in \mathbb{N}^*$.

If $f(1) > 1$, then by (6) we have, $\forall n \in \mathbb{N}^*$,

$$f(4) - f(2) = f(2n-4) - f(2n+2) = 0,$$

and then

$$f(2n) = f(2). \quad (7)$$

Using (2), (5) and (7), we have $\{f(1), f(2)\} = \{2, 1998\}$, and hence,

$$f(n) = \begin{cases} 2 & \text{if } n \text{ odd,} \\ 1998 & \text{if } n \text{ even;} \end{cases} \quad \text{or} \quad f(n) = \begin{cases} 1998 & \text{if } n \text{ odd,} \\ 2 & \text{if } n \text{ even.} \end{cases}$$

Case 2: $f(3) > f(1)$.

By (3) we have $f(2n-1) < f(2n+1)$ for all $n \in \mathbb{N}^*$. Now, by (4) we have

$$f(4) - f(2) = f(2n+4) - f(2n+2) = 0, \quad \forall n \in \mathbb{N}^*.$$

Then

$$f(2n) = f(2) \quad \forall n \in \mathbb{N}^*, \tag{8}$$

and

$$f(3) - f(1) = (f(2))^n (f(2n+3) - f(2n+1)) \quad \forall n \in \mathbb{N}^*.$$

Because $f(2n+3) - f(2n+1) > 0$, we see that $f(2) > 1$ is impossible. Thus, $f(2) = 1$ and $f(3) - f(1) = f(2n+3) - f(2n+1)$ for all positive n. Now, from (2), we obtain that $f(3) - f(1) = 1997$, and f is given by

$$f(n) = \begin{cases} 1 & \text{if } n \text{ even,} \\ \left(a + \frac{n-1}{2}\right) 1997 & \text{if } n \text{ odd,} \end{cases}, \quad \text{where } a \in \mathbb{N}^*.$$

Vietnamese Mathematical Olympiad, 1996 (Category B)

Determine all functions $f : \mathbb{Z} \to \mathbb{Z}$ satisfying simultaneously two conditions:

(i) $f(1995) = 1996$

(ii) for every $n \in \mathbb{Z}$, if $f(n) = m$, then $f(m) = n$ and $f(m+3) = n - 3$, ($\mathbb{Z}$ is the set of integers).

Solution

Let $f : \mathbb{Z} \to \mathbb{Z}$ satisfy the two conditions. From (ii), we deduce: $f(f(n)) = n$ for all $n \in \mathbb{Z}$; that is, f is involutory.

The equality $f(m+3) = n-3$ then gives $m+3 = f(n-3)$ or $f(n-3) = f(n) + 3$. From this, we easily obtain, by induction, $f(n - 3k) = f(n) + 3k$ for all positive integers k.

Let m be any integer; m may be written $m = f(n)$ for an n (for $n = f(m)$, actually). Condition (ii) gives $f(m + 3) = f(m) - 3$, and by induction: $f(m + 3k) = f(m) - 3k$ for all positive integers k. All this can be summed up by $f(n + 3k) = f(n) - 3k$ for all $n, k \in \mathbb{Z}$.

It follows that for every $k \in \mathbb{Z}$,

$$\begin{aligned} f(3k) &= f(0) - 3k, \\ f(3k+1) &= f(1) - 3k, \\ f(3k+2) &= f(2) - 3k. \end{aligned}$$

Let $k = 665$. We have $f(1995) = f(0) - 1995 = 1996$ and $f(1996) = 1995 = f(1) - 1995$. Then $f(0) = 3991$ and $f(1) = 3990$.

Thus, for every $k \in \mathbb{Z}$,

$$\begin{aligned} f(3k) &= 3991 - 3k, \\ f(3k+1) &= 3990 - 3k, \\ f(3k+2) &= f(2) - 3k. \end{aligned}$$

We remark that

$$\begin{aligned} f(3\mathbb{Z}) &= 3\mathbb{Z} + 1, \\ f(3\mathbb{Z}+1) &= 3\mathbb{Z}, \\ \text{and so,} \quad f(3\mathbb{Z}+2) &\subset 3\mathbb{Z} + 2. \end{aligned}$$

Moreover f is bijective, so $f(3\mathbb{Z} + 2) = 3\mathbb{Z} + 2$. Denote $f(2) = 3a + 2$, where $a \in \mathbb{Z}$. Then:

if f is a solution then $f(m) = \begin{cases} 3991 - m & \text{if } m \not\equiv 2 \pmod{3} \\ 3a + 4 - m & \text{if } m \equiv 2 \pmod{3}. \end{cases}$

Conversely, a straightforward calculation shows that any such f satisfies the conditions.

19th Austrian-Polish Mathematics Competition, 1996

The polynomials $P_n(x)$ are defined recursively by $P_0(x) = 0$, $P_1(x) = x$ and
$$P_n(x) = xP_{n-1}(x) + (1-x)P_{n-2}(x) \quad \text{for} \quad n \geq 2.$$
For every natural number $n \geq 1$, find all real numbers x satisfying the equation $P_n(x) = 0$.

Solution

We will prove that for $n \geq 1$, the only real solution of $P_n(x) = 0$ is $x = 0$.

For $n \geq 2$,
$$P_n(x) - P_{n-1}(x) = (x-1)(P_{n-1}(x) - P_{n-2}(x)).$$
Then an easy induction leads to
$$P_n(x) - P_{n-1}(x) = (x-1)^{n-1}(P_1(x) - P_0(x)) = x(x-1)^{n-1}.$$
That is, $P_n(x) = P_{n-1}(x) + x(x-1)^{n-1}$ for $n \geq 2$, and we note that it remains true for $n = 1$.

We deduce that, for $n \geq 1$,
$$\begin{aligned} P_n(x) &= x(x-1)^{n-1} + x(x-1)^{n-2} + \cdots + x + P_0(x) \\ &= x((x-1)^{n-1} + (x-1)^{n-2} + \cdots + 1). \end{aligned}$$

Then, if $x = 2$, $P_n(2) = 2n \neq 0$ and if $x \neq 2$, $P_n(x) = x \cdot \frac{((x-1)^n - 1)}{x-2}$. Thus, $P_n(x) = 0$ if and only if $x = 0$ or $(x-1)^n = 1$ for $x \neq 2$.

If n is even, $(x-1)^n = 1$ if and only if $x = 0$ or $x = 2$. Then $P_n(x) = 0$ if and only if $x = 0$.

If n is odd, $(x-1)^n = 1$ if and only if $x = 2$. Then $P_n(x) = 0$ if and only if $x = 0$.

Thus, for $n \geq 1$, $P_n(x) = 0$ if and only if $x = 0$.

3rd Turkish Mathematical Olympiad, 1995

Let $\mathbb{N}$ denote the set of positive integers. Let A be a real number and $(a_n)_{n=1}^{\infty}$ be a sequence of real numbers such that $a_1 = 1$ and

$$1 < \frac{a_{n+1}}{a_n} \leq A \quad \text{for all} \quad n \in \mathbb{N}.$$

(a) Show that there is a unique non-decreasing surjective function $k : \mathbb{N} \to \mathbb{N}$ such that $1 < \frac{A^{k(n)}}{a_n} \leq A$ for all $n \in \mathbb{N}$.

(b) If k takes every value at most m times, show that there exists a real number $C > 1$ such that $C^n \leq Aa_n$ for all $n \in \mathbb{N}$.

Solution

(a) The condition $1 < \frac{A^{k(n)}}{a_n} \leq A$ is equivalent to $A^{k(n)-1} \leq a_n < A^{k(n)}$ or $k(n) - 1 \leq \frac{\ln(a_n)}{\ln(A)} < k(n)$. Hence, k is necessarily the function given by $k(n) = 1 + \left\lfloor \frac{\ln(a_n)}{\ln(A)} \right\rfloor = 1 + \lfloor \log_A(a_n) \rfloor$ for all $n \in \mathbb{N}$. This shows the unicity of k. Since $\{a_n\}$ is increasing and $a_1 = 1$, we have $a_n \geq 1$. We also note $A > 1$. Hence, $\frac{\ln(a_n)}{\ln(A)}$ is a non-negative real number and $1 + \lfloor \log_A(a_n) \rfloor$ is a positive integer. Thus, we can define a function $k : \mathbb{N} \to \mathbb{N}$ by the formula $k(n) = 1 + \lfloor \log_A(a_n) \rfloor$. For all $n \in \mathbb{N}$: $a_n < a_{n+1}$, so that $\log_A(a_n) < \log_A(a_{n+1})$ and $k(n) \leq k(n+1)$. Thus, k is non-decreasing. Now, let us remark that, for $s \in \mathbb{N}$, $k(n) = s$ is equivalent to $A^{s-1} \leq a_n < A^s$. Therefore, if $\{a_n\}$ is bounded above, say $a_n \leq M$ for all n, an s satisfying $A^{s-1} > M$ (such an s exists since $A > 1$) cannot be an image under k. Thus, if $\{a_n\}$ satisfies the hypothesis and is convergent (for instance $a_n = \frac{1}{2}\left(1 + \frac{1}{n}\right)^n$), the surjectivity of k cannot be obtained. Consequently, we will henceforth assume that $\{a_n\}$ is not bounded above.

Given $s \in \mathbb{N}$, we prove that the equation $k(n) = s$ has at least one solution. For $s = 1$, $n = 1$ is obviously a solution, so now we suppose that $s \geq 2$. From the supplementary hypothesis, there exists $n \in \mathbb{N}$ such that $a_n \geq A^{s-1}$. Let r be the least of these n's, so that $a_{r-1} < A^{s-1} \leq a_r$. Then $a_r \leq Aa_{r-1} < A^s$ so that $A^{s-1} \leq a_r < A^s$ and r is a solution. The function k fulfils all the demands and (a) follows.

(b) For each $s \in \mathbb{N}$, denote by $N(s)$ the number of n such that $k(n) = s$; that is, such that $A^{s-1} \leq a_n < A^s$. By hypothesis, we have $N(s) \leq m$ for each s.

We will show that (b) holds with $C = A^{1/m} > 1$. Let n be an arbitrary integer. If $k(n) = s$, then

$$n = N(1) + N(2) + \cdots + N(s-1) + j \quad \text{where} \quad j \in \{1, 2, \ldots, N(s)\}$$

(because there are $N(1) + N(2) + \cdots + N(s-1)$ terms of the sequence $\{a_n\}$ which are $< A^{s-1}$). On the one hand, $a_n A \geq A^{s-1}A = A^s = C^{ms}$, and, on the other hand,

$$\begin{aligned} n &= N(1) + N(2) + \cdots + N(s-1) + j \\ &\leq N(1) + N(2) + \cdots + N(s-1) + N(s) \leq sm. \end{aligned}$$

Hence, $C^n \leq C^{ms}$ and we obtain $C^n \leq a_n A$. So (b) is proved.

3rd Turkish Mathematical Olympiad, 1995

Let $\mathbb{N}$ denote the set of positive integers. Find all surjective functions $f : \mathbb{N} \to \mathbb{N}$ satisfying the condition
$$m \mid n \iff f(m) \mid f(n)$$
for all $m, n \in \mathbb{N}$.

Solution

For such a function f, we have:

(a) $f(1) = 1$: Since $1 \mid n$, we have $f(1) \mid f(n)$ for all $n \in \mathbb{N}$. But f is surjective, so that $f(n)$ may be any positive integer; hence, $f(1)$ divides all positive integers and $f(1) = 1$.

(b) f is bijective: f is already surjective; moreover, if $f(m) = f(n)$, then $f(m) \mid f(n)$ and $f(n) \mid f(m)$ so that $m \mid n$ and $n \mid m$; that is, $m = n$. Hence, f is also injective.

Obviously, the condition on f also reads: $m \mid n \iff f^{-1}(m) \mid f^{-1}(n)$ for all $m, n \in \mathbb{N}$.

(c) $f(p)$ is prime whenever p is prime: If k divides $f(p)$, $f^{-1}(k)$ divides p, so that $f^{-1}(k) = 1$ or p and $k = 1$ or $f(p)$.

More generally, $f(p^a) = (f(p))^a$ for all positive integers a: Let q be a prime integer dividing $f(p^a)$. Then $f^{-1}(q)$ is a prime integer dividing p^a. It follows that $f^{-1}(q) = p$ and $q = f(p)$. Therefore, $f(p^a)$ is a power of $f(p)$, say $f(p^a) = (f(p))^b$. Now, the $a + 1$ distinct divisors of p^a, namely $1, p, \ldots, p^a$, provide $a + 1$ distinct divisors of $f(p^a) = (f(p))^b$, namely $1, f(p), \ldots, f(p^a)$. As $(f(p))^b$ has $b + 1$ divisors, we have $a \leq b$. Similarly, using f^{-1}, $b \leq a$. Thus, $a = b$ and $f(p^a) = (f(p))^a$.

(d) If u, v are coprime positive integers, then $f(u)$, $f(v)$ are also coprime and $f(uv) = f(u)f(v)$. That $f(u)$, $f(v)$ are coprime is immediate since any prime integer p dividing $f(u)$ and $f(v)$ would provide a prime integer, namely $f^{-1}(p)$, dividing u and v.

Since $u \mid uv$ and $v \mid uv$, $f(u) \mid f(uv)$ and $f(v) \mid f(uv)$. Taking into account the result we have just shown, we get $f(u) \cdot f(v) \mid f(uv)$. With f^{-1}, $f(u)$, $f(v)$ instead of f, u, v, we obtain $u \cdot v \mid f^{-1}(f(u) \cdot f(v))$ and consequently $f(uv) \mid f(u) \cdot f(v)$. Hence, $f(uv) = f(u)f(v)$. Let P be the set of prime integers. We have obtained the following concerning f: The restriction of f to P is a bijection from P onto P and, combining the last two results, if $n = \prod_{p \in P} p^{v_p(n)}$, then $f(n) = \prod_{p \in P} (f(p))^{v_p(n)}$. [Here, $v_p(n)$ denotes the exponent of p in the standard factorization of n into prime powers: of course, we have $v_p(n) = 0$ for all but a finite number of p and the sequence $(v_p(n))_{p \in P}$ is uniquely determined by n.]

Conversely, let $\phi : P \to P$ be any bijection from P onto P and let $f : \mathbb{N} \to \mathbb{N}$ be defined by

$$f(n) = \prod_{p \in P} (\phi(p))^{v_p(n)} \quad \text{when} \quad n = \prod_{p \in P} p^{v_p(n)}. \tag{1}$$

Clearly, f is surjective. Moreover,

$$\begin{aligned} m \mid n &\iff v_p(m) \leq v_p(n) && \text{for all } p \in P \\ &\iff v_{\phi(p)}(f(m)) \leq v_{\phi(p)}(f(n)) && \text{for all } p \in P \\ &\iff v_q(f(m)) \leq v_q(f(n)) && \text{for all } q \in P \\ &\iff f(m) \mid f(n). \end{aligned}$$

In conclusion, the solutions are the functions extending to $\mathbb{N}$ the bijections $\phi : P \to P$ by means of the formula (1).

Australian Mathematical Olympiad, 1996

Let $p(x)$ be a cubic polynomial with roots r_1, r_2, r_3. Suppose that

$$\frac{p\left(\frac{1}{2}\right) + p\left(-\frac{1}{2}\right)}{p(0)} = 1000.$$

Find the value of $\frac{1}{r_1 r_2} + \frac{1}{r_2 r_3} + \frac{1}{r_3 r_1}$.

Solution

Note that $p(0)$ is supposed non-zero, so that r_1, r_2, r_3 are non-zero. Let $p(x) = ax^3 + bx^2 + cx + d$.

The hypothesis is: $\left(\frac{a}{8} + \frac{b}{4} + \frac{c}{2} + d\right) + \left(-\frac{a}{8} + \frac{b}{4} - \frac{c}{2} + d\right) = 1000d$; that is, $b = 1996d$. Now,

$$\frac{1}{r_1 r_2} + \frac{1}{r_2 r_3} + \frac{1}{r_3 r_1} = \frac{r_3 + r_1 + r_2}{r_1 r_2 r_3} = \frac{-b/a}{-d/a} = \frac{b}{d} = 1996.$$

Australian Mathematical Olympiad, 1996

Let f be a function that is defined for all integers and takes only the values 0 and 1. Suppose f has the following properties:

(i) $f(n+1996) = f(n)$ for all integers n;

(ii) $f(1) + f(2) + \cdots + f(1996) = 45$.

Prove that there exists an integer t such that $f(n+t) = 0$ for all n for which $f(n) = 1$ holds.

Solution

Let C be a circle with radius $\frac{1996}{2\pi}$. The real line can be wrapped around the circle in the counterclockwise sense so that the integer i is mapped to the point A_i of the circle.

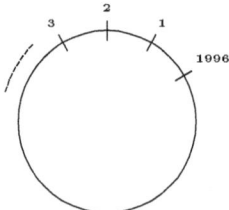

Since the circumference of the circle is 1996, if two numbers differ by 1996, they map to the same point of the circle, (that is, $A_{i+1996} = A_0$ for $i \in \mathbb{Z}$.)

From (i), the integers are represented on C by the points $A_1, \ldots, A_{1996}$, and we may colour those A_i such that $f(i) = 1$. By (ii), and since $f(n) \in \{0, 1\}$, there are exactly 45 coloured points.

If $i < j$ and $i, j \in \{1, \ldots, 1996\}$, A_i and A_j determine two arcs of lengths $j - i$ and $i - j + 1996$ (which may be equal). The possible lengths are $1, 2, 3, \ldots, 1995$. But there are 45 coloured points, and distinct coloured points determine at most $2\binom{45}{2} = 1980$ different lengths.

Then, for all i, there is $t \in \{1, \ldots, 1995\}$ such that if A_i is coloured then A_{i+t} is not coloured (with subscripts read modulo 1996).

That is, for all i, if $f(i) = 1$, then $f(i+t) = 0$.

47th Polish Mathematical Olympiad, 1995

Find all pairs (n, r), with n a positive integer, r a real number, for which the polynomial $(x+1)^n - r$ is divisible by $2x^2 + 2x + 1$.

Solution

By division, we can find a polynomial $q(x)$ and two real numbers a, b such that:

$$(x+1)^n - r = (2x^2 + 2x + 1) \cdot q(x) + ax + b.$$

Since the complex roots of $2x^2 + 2x + 1$ are $-\frac{1}{2} + \frac{i}{2}$ and $-\frac{1}{2} - \frac{i}{2}$, the numbers a, b satisfy the system:

$$\begin{cases} a\left(-\frac{1}{2} + \frac{i}{2}\right) + b = \left(1 + \frac{-1+i}{2}\right)^n - r \\ a\left(-\frac{1}{2} - \frac{i}{2}\right) + b = \left(1 + \frac{-1-i}{2}\right)^n - r. \end{cases}$$

Taking into account:

$$1 + \frac{-1+i}{2} = \frac{1+i}{2} = \frac{1}{\sqrt{2}} e^{i\pi/4}$$

and

$$1 + \frac{-1-i}{2} = \frac{1-i}{2} = \frac{1}{\sqrt{2}} e^{-i\pi/4},$$

we get, by subtraction:

$$ai = 2^{-n/2}(e^{in\pi/4} - e^{-in\pi/4}) = 2^{-n/2} \cdot 2i \sin(n\pi/4).$$

Hence, $a = 0$ if and only if $\sin(n\pi/4) = 0$; that is, n is a multiple of 4. Assuming that $n = 4k$, where k is a positive integer, we obtain that $b = 2^{-2k}(e^{i\pi})^k - r$, so that $b = 0$ if and only if $r = \frac{(-1)^k}{4^k}$. Thus, the solutions are the pairs $\left(4k, \frac{(-1)^k}{4^k}\right)$ where k is a positive integer.

10th Nordic Mathematical Contest, 1996

A real-valued function f is defined for positive integers, and a positive integer a satisfies

$$f(a) = f(1995), \quad f(a+1) = f(1996), \quad f(a+2) = f(1997),$$

$$f(n+a) = \frac{f(n)-1}{f(n)+1} \quad \text{for any positive integer } n.$$

(a) Prove that $f(n+4a) = f(n)$ for any positive integer n.
(b) Determine the smallest possible value of a.

Solution

(a) From $f(n+a) = \frac{f(n)-1}{f(n)+1}$, we deduce that

$$f(n+2a) = f((n+a)+a) = \frac{\frac{f(n)-1}{f(n)+1} - 1}{\frac{f(n)-1}{f(n)+1} + 1} = -\frac{1}{f(n)} \quad \text{and}$$

$$f(n+4a) = f((n+2a)+2a) = -\frac{1}{f(n+2a)} = f(n).$$

(b) The smallest possible value of a is 3. Indeed, if $a = 1$, then

$$f(1) = f(a) = f(1995) = f(3+498\cdot 4a) = f(3) = f(1+2a) = -\frac{1}{f(1)},$$

and hence,

$$(f(1))^2 = -1, \quad \text{which is impossible.}$$

If $a = 2$, then

$$\begin{aligned}f(2) &= f(a) = f(1995) = f(3+249\cdot 4a) = f(3) = f(a+1) \\ &= f(1996) = f(4+249\cdot 4a) = f(4) = f(2+a) = \frac{f(2)-1}{f(2)+1},\end{aligned}$$

and hence, $(f(2))^2 = -1$, which is impossible.

If $a = 3$, we should be sure that $f(n) \neq -1$ for all $n \in \mathbb{N}$. Let $f(1)$, $f(2)$ and $f(3)$ be chosen arbitrarily different from $-1, 0, 1$. Then, for $n = 1$, 2 or 3, and since $f(n) \neq -1, 0, 1$, none of $f(n)$, $f(n+3) = \frac{f(n)-1}{f(n)+1}$, $f(n+6) = -\frac{1}{f(n)}$ and $f(n+9) = -\frac{f(n)+1}{f(n)-1}$ is equal to -1, and then, by (a), $f(n) \neq -1$ for all $n \in \mathbb{N}$.

By construction, we have

$$f(n+a) = f(n+3) = \frac{f(n)-1}{f(n)+1},$$

and, by (a),

$$f(n+12) = f(n+4a) = f(n).$$

Whence,

$$f(a) = f(3) = f(3+166\cdot 12) = f(1995),$$
$$f(a+1) = f(4) = f(4+166\cdot 12) = f(1996),$$
$$f(a+2) = f(5) = f(5+166\cdot 12) = f(1997),$$

as required.

9th Irish Mathematical Olympiad, 1996

For each positive integer n, let $f(n)$ denote the greatest common divisor of $n!+1$ and $(n+1)!$ (where ! denotes "factorial"). Find, with proof, a formula for $f(n)$ for each n.

Solution

We show that
$$f(n) = \begin{cases} n+1 & \text{if } n+1 \text{ is a prime}, \\ 1 & \text{otherwise}. \end{cases}$$

For convenience of notation, denote $f(n)$ by d. Since $d \mid n!+1$ and $d \mid (n+1)!$ we have $d \mid (n+1)(n!+1) - (n+1)!$; that is, $d \mid n+1$. If $n+1$ is a prime, then $n+1 \mid n!+1$ by Wilson's Theorem. Since clearly $n+1 \mid (n+1)!$ we have $n+1 \mid d$. Hence, $d = n+1$. If $n+1$ is a composite, then $n+1 = ab$ for some integers a and b such that $1 < a \leq b < n$. If $d = n+1$ then $ab = d$ and so, $a \mid d$. Since $d \mid n!+1$ we have $a \mid n!+1$. On the other hand, since $a < n$ we also have $a \mid n!$. Hence, $a \mid 1$ which implies that $a = 1$, a contradiction. Thus, $d \leq n$. Then, $d \mid n!$ together with $d \mid n!+1$ imply that $d = 1$, and the proof is complete.

9th Irish Mathematical Olympiad, 1996

Let K be the set of all real numbers x with $0 \leq x \leq 1$. Let f be a function from K to the set of all real numbers $\mathbb{R}$ with the following properties:

(i) $f(1) = 1$.

(ii) $f(x) \geq 0$ for all $x \in K$.

(iii) if x, y and $x + y$ are all in K, then
$$f(x+y) \geq f(x) + f(y).$$

Prove that $f(x) \leq 2x$ for all $x \in K$.

Solution

We first prove the following lemma.

Lemma. If $0 \leq x \leq \frac{1}{n}$ for $n \in \mathbb{N}$, then $f(x) \leq \frac{1}{n}$.

Proof of Lemma. Let $0 \leq x \leq \frac{1}{n}$ for $n \in \mathbb{N}$. Then, we have
$$1 = f(1) = f(1 - nx + nx) \geq f(1-nx) + f(nx) \geq f(nx),$$
and we have
$$f(nx) = f(\overbrace{x + \cdots + x}^{n \text{ times}}) \geq \overbrace{f(x) + \cdots + f(x)}^{n \text{ times}}$$
from (iii).

Hence, we get
$$1 \geq f(nx) \geq nf(x) \quad \text{or} \quad \frac{1}{n} \geq f(x),$$
as desired.

We shall prove that $f(x) \leq 2x$ for $0 < x \leq 1$.

Let $0 < x \leq 1$. Then, there exists a natural number x such that $\frac{1}{n+1} < x \leq \frac{1}{n}$. Then we have $f(x) \leq \frac{1}{n}$ from the above lemma.

So, we have $f(x) \leq \frac{1}{n} \leq \frac{2}{n+1} < 2x$ or $f(x) < 2x$, as desired.

Now, we prove $f(0) \leq 0$.

Since $0 \leq \frac{1}{n}$ for all $n \in \mathbb{N}$, we have $f(0) \leq \frac{1}{n}$ for all $n \in \mathbb{N}$ from the above lemma. This implies that $f(0) \leq 0$ since $\lim_{n \to \infty} \frac{1}{n} = 0$.

St. Petersburg City Mathematical Olympiad, 1996 (Third Round)

Find all quadruplets of polynomials $p_1(x), p_2(x), p_3(x), p_4(x)$ with real coefficients possessing the following remarkable property: for all integers x, y, z, t satisfying the condition $xy - zt = 1$, the equality $p_1(x)p_2(y) - p_3(z)p_4(t) = 1$ holds.

Solution

If p_1, p_2, p_3, p_4 are such polynomials, then for all $x \in \mathbb{Z}$:

$$p_1(x+1)p_2(1) - p_3(x)p_4(1) = 1 \qquad (1)$$

and

$$p_1(x+1)p_2(1) - p_3(1)p_4(x) = 1. \qquad (2)$$

Thus,

$$p_3(x)p_4(1) = p_3(1)p_4(x). \qquad (3)$$

Moreover,

$$p_1(x)p_2(1) - p_3(x-1)p_4(1) = 1$$

and

$$p_1(1)p_2(x) - p_3(x-1)p_4(1) = 1.$$

Thus,

$$p_1(x)p_2(1) = p_1(1)p_2(x). \qquad (4)$$

Moreover:

$$p_1(x)p_2(x) - p_3(x-1)p_4(x+1) = 1 \qquad (5)$$

and

$$p_1(x^2)p_2(1) - p_3(x-1)p_4(x+1) = 1.$$

Thus,

$$p_1(x^2)p_2(1) = p_1(x)p_2(x). \qquad (6)$$

Since (3), (4), (5), (6) hold for an infinite number of values, and since they are relations between polynomials, then they hold for all real numbers x.

Case 1: If $p_1 \equiv 0$, then from (1):

$$p_3(x)p_4(1) = -1 \quad \text{for all } x \in \mathbb{R}.$$

It follows that p_3 is constant.

Since p_3 and p_4 are playing symmetric parts, we also have p_4 is constant. Then $p_3(x) = c$ and $p_4(x) = -\frac{1}{c}$ where $c \in \mathbb{R}^*$.

Conversely: $(0, p_2, c, -\frac{1}{c})$, where $p_2 \in \mathbb{R}[x]$ is arbitrary and $c \in \mathbb{R}^*$, is a solution of the problem.

By the same reasoning:

- $p_2 \equiv 0$ gives the solutions $(p_1, 0, c, -\frac{1}{c})$ where p_1 is an arbitrary polynomial, $p_1 \in \mathbb{R}[x]$, and $c \in \mathbb{R}^*$.

- $p_3 \equiv 0$ gives the solutions $(a, \frac{1}{a}, 0, p_4)$.

- $p_4 \equiv 0$ gives the solutions $(a, \frac{1}{a}, p_3, 0)$ where $a \in \mathbb{R}^*$ and p_3 (resp. p_4) is an arbitrary polynomial.

Now, we suppose that no one of p_1, p_2, p_3, p_4 is identically zero.

Case 2: If $p_1 \equiv a$ where $a \in \mathbb{R}^*$, then from (4), we have $p_2(x) = b$ where $b \in \mathbb{R}^*$.

If $p_3(1) = 0$, then, from (2) we have $ab = 1$. Thus, from (5), we have $p_3(x-1)p_4(x+1) = 0$ for all $x \in \mathbb{Z}$.

It follows that at least one of the polynomials p_3, p_4 has an infinite number of zeros. Thus, $p_3 \equiv 0$ or $p_4 \equiv 0$. Contradiction. We deduce that $p_3(1) \neq 0$.

Then, from (2), p_4 is constant: $p_4 \equiv d$, where $d \in \mathbb{R}^*$. In the same way, $p_3 \equiv c$, where $c \in \mathbb{R}^*$. And we must have $ab - cd = 1$.

Conversely: (a, b, c, d), where $a, b, c, d \in \mathbb{R}^*$ and $ab - cd = 1$, is a solution.

Notice that the condition a, b, c, d non-zero can be eased if $ab - cd = 1$ because it gives solutions of the first case.

Moreover, by the same reasoning, the cases p_3 is constant, p_4 is constant, p_2 is constant, give the same solutions.

Now, we suppose that:

Case 3: No one of p_1, p_2, p_3, p_4 is constant.

Then, for all $x \in \mathbb{Z}$: $p_1(x)p_2(0) - p_3(-1)p_4(1) = 1$ where p_1 is a non-constant polynomial. Thus, $p_2(0) = 0$.

In the same way, $p_1(0) = p_3(0) = p_4(0) = 0$. Then, we have $p_1(1)p_2(1) = p_1(1)p_2(1) - p_3(0)p_4(0) = 1$. It follows that $p_1(1) \neq 0$ and $p_2(1) \neq 0$.

In the same way, $-p_3(1)p_4(-1) = p_1(0)p_2(0) - p_3(1)p_4(-1) = 1$. Then $p_3(1) \neq 0$, $p_4(-1) \neq 0$.

But, it is clear that if (p_1, p_2, p_3, p_4) is a solution then $(ap_1, \frac{1}{a}p_2, bp_3, \frac{1}{b}p_4)$ is also a solution where a, b are arbitrary non-zero real numbers.

Then, with no loss of generality, we suppose that $p_1(1) = p_3(1) = 1$. Thus, $p_2(1) = 1$, $p_4(-1) = -1$.

From (4) and (6), we have $p_1(x^2) = p_1^2(x)$ for all $x \in \mathbb{R}$. We let $p_1(x) = \sum_{i=0}^{n} a_i x^i$, $n \in \mathbb{N}^*$.

Identifying the coefficients, we have:

- $a_n = a_n^2$ with $a_n \neq 0$, then $a_n = 1$.
- $2a_n a_{n-1} = 0$, then $a_{n-1} = 0$.
- $2a_n a_{n-2} + a_{n-1}^2 = a_{n-1}$.

From the above, we have $a_{n-2} = 0$. And so on ... an easy induction leads to $a_k = 0$ for all $k < n$. Then $p_1(x) = x^n$ for some $n \in \mathbb{N}^*$.

From (4), we deduce that $p_2(x) = x^n$.

From (1), we have
$$(x+1)^n - 1 = p_3(x)p_4(1) = p_4(x)p_3(1) = p_4(x).$$

Then,
$$p_4(x) = (x+1)^n - 1$$
and
$$p_3(x) = \frac{(x+1)^n - 1}{2^n - 1} \quad \text{(since } p_3(1) = 1\text{)}.$$

Then, for all x, y, z, t integers such that $xy - zt = 1$, we have
$$(xy)^n - \frac{((z+1)^n - 1)((t+1)^n - 1)}{2^n - 1} = 1.$$

Thus, for all $t \in \mathbb{Z}$, with $z = -1$, $x = 1$, $y = 1 - t$, we must have
$$(1-t)^n + \frac{(1+t)^n - 1}{2^n - 1} = 1.$$

It follows that the polynomial $Q(t) = (1-t)^n + \frac{(1+t)^n - 1}{2^n - 1}$ is constant. Since $n \geq 1$, the coefficient of t^n must be zero. That is,
$$(-1)^n + \frac{1}{2^n - 1} = 0.$$

Then,
$$2^n - 1 = 1 \quad \text{or} \quad 2^n - 1 = -1.$$

Thus, $n = 1$. It follows that
$$p_1(x) = p_2(x) = p_3(x) = p_4(x) = x.$$

Conversely: (x, x, x, x) is obviously a solution.

Conversely: The solutions are the quadruplets of the form:

(a) $\left(0, P, a, -\frac{1}{a}\right)$, $\left(P, 0, a, -\frac{1}{a}\right)$, $\left(a, \frac{1}{a}, P, 0\right)$, $\left(a, \frac{1}{a}, 0, P\right)$, where $P \in \mathbb{R}[x]$ and $a \in \mathbb{R}^*$ are arbitrary.

(b) (a, b, c, d), where $a, b, c, d \in \mathbb{R}$ with $ab - cd = 1$. This is, therefore, $\left(ax, \frac{1}{a}x, bx, \frac{1}{b}x\right)$ where $a, b \in \mathbb{R}^*$ are arbitrary.

Republic of Moldova XL Mathematical Olympiad, 1996 (11-12) 5

Let p be the number of functions defined on the set $\{1, 2, \ldots, m\}$, $m \in N^*$, with values in the set $\{1, 2, \ldots, 35, 36\}$ and q be the number of functions defined on the set $\{1, 2, \ldots, n\}$, $n \in N^*$, with values in the set $\{1, 2, 3, 4, 5\}$. Find the least possible value for the expression $|p - q|$.

Solution

Let $m, n \in N^*$. We have $p = 36^m$ and $q = 5^n$. The problem is then to find the least possible value of $|36^m - 5^n|$ over all $m, n \in N^*$.

For $m, n \in N^*$,

$$36^m = a \pmod{100} \quad \text{where } a \in \{36, 96, 56, 16, 76\},$$
$$5^n = 25 \pmod{100} \quad \text{for } n \geq 2.$$

Since $36 - 5^2 = 11$, the least possible value is then 9 or 11. But

$$36^m - 5^n = -5^n \pmod 9 \neq 0 \pmod 9,$$

so that $36^m - 5^n = \pm 9$ is impossible.

It follows that the least possible value of $|36^m - 5^n|$ is 11.

Ukrainian Mathematical Olympiad, 1996

(11th grade) Does a function $f : \mathbb{R} \to \mathbb{R}$ exist which is not a polynomial and such that for all real x

$$(x-1)f(x+1) - (x+1)f(x-1) = 4x(x^2-1)?$$

Solution

Yes! Take any function $f : x \mapsto x^3 + xk(x)$ where $k : \mathbb{R} \to \mathbb{R}$ is a bounded, non-constant, 2–periodic function (for example, $k(x) = \sin(\pi x)$, $k(x) = x - \lfloor x \rfloor, \dots$). For such an f and for all real numbers x:

$$
\begin{aligned}
&(x-1)f(x+1) - (x+1)f(x-1) \\
&= (x-1)(x+1)^3 - (x+1)(x-1)^3 + (x^2-1)(k(x+1) - k(x-1)) \\
&= 4x(x^2-1) + 0 \quad \text{(because of the periodicity of } k\text{)} \\
&= 4x(x^2-1).
\end{aligned}
$$

Moreover f is clearly not a polynomial: if it were, it would be a multiple of x (since $f(0) = 0$) and $k(x)$ would be a polynomial as well, which is impossible since k is non-constant and bounded.

Taiwan Mathematical Olympiad, 1996

Let $q_0, q_1, q_2, \ldots$ be a sequence of integers such that

(a) for any $m > n$, $m - n$ is a factor of $q_m - q_n$, and

(b) $|q_n| \leq n^{10}$ for all integers $n \geq 0$.

Show that there exists a polynomial $Q(x)$ satisfying $Q(n) = q_n$ for all n.

Solution

This is a particular case of problem 4 of the 1995 USAMO, where it was asked:

"Suppose $q_0, q_1, \ldots$ is an infinite sequence of integers satisfying the following two conditions:

(i) $m - n$ divides $q_m - q_n$ for $m > n \geq 0$.

(ii) there is a polynomial P such that $|q_n| < P(n)$ for all n.

Show there is a polynomial Q such that $q_n = Q(n)$ for all n."

Croatian Mathematical Olympiad, 1995 (IV Class)

Determine all functions $f : \mathbb{R} \longrightarrow \mathbb{R}$ continuous at 0, which satisfy the following relation

$$f(x) - 2f(tx) + f(t^2x) = x^2 \quad \text{for all} \quad x \in \mathbb{R},$$

where $t \in (0, 1)$ is a given number.

Solution

First, we introduce the function $g(x) = f(x) - f(tx)$ defined on $\mathbb{R}$. Then g is also continuous at 0.

Since $g(tx) = f(tx) - f(t^2x)$, we easily get $g(x) - g(tx) = x^2$. Therefore, we have

$$g(x) - g(t^{n+1}x) = \sum_{i=1}^{n+1} \{g(t^{i-1}x) - g(t^i x)\} = \sum_{i=1}^{n+1} (t^{i-1}x)^2,$$

or

$$g(x) = g(t^{n+1}x) + \left(\sum_{i=1}^{n+1} (t^2)^{i-1}\right) x^2.$$

We note that $\lim_{n \to \infty} t^{n+1} = 0$ since $t \in (0, 1)$. By the continuity of g at 0, we have

$$g(x) = \lim_{n \to \infty} g(t^{n+1}x) + \lim_{n \to \infty} \left(\sum_{i=1}^{n+1} (t^2)^{i-1}\right) x^2$$
$$= g(0) + \frac{1}{1-t^2} x^2.$$

Since $g(0) = f(0) - f(t \cdot 0) = 0$, we obtain $g(x) = \frac{1}{1-t^2} x^2$. Thus, we have $f(x) - f(tx) = \frac{1}{1-t^2} x^2$.

Also, we get

$$f(x) - f(t^{n+1}x) = \sum_{i=1}^{n+1} \{f(t^{i-1}x) - f(t^i x)\} = \frac{1}{1-t^2} \sum_{i=1}^{n+1} (t^{i-1}x)^2.$$

By the continuity of f at 0, we have

$$f(x) = \lim_{n \to \infty} f(t^{n+1}x) + \frac{1}{1-t^2} \lim_{n \to \infty} \sum_{i=1}^{n+1} (t^{i-1}x)^2$$

or

$$f(x) = f(0) + \frac{1}{(1-t^2)^2} x^2.$$

Therefore, $f(x) = \frac{1}{(1-t^2)^2} x^2 + C$, $(C \in \mathbb{R})$.

13th Iranian Mathematical Olympiad 1995

Let $P(x)$ be a polynomial with rational coefficients such that $P^{-1}(\mathbb{Q}) \subseteq \mathbb{Q}$. Show that P is linear.

Solution

Let $P(x) \in \mathbb{Q}[x]$ such that $P^{-1}(\mathbb{Q}) \subset \mathbb{Q}$. It follows that

$$P(\mathbb{Q}) \subset \mathbb{Q} \text{ (since } P \in \mathbb{Q}[x]) \tag{1}$$

and

$$P(\mathbb{R} - \mathbb{Q}) \subset \mathbb{R}\backslash\mathbb{Q} \text{ (since } P^{-1}(\mathbb{Q}) \subset \mathbb{Q}) . \tag{2}$$

It is easy to see that P cannot be a constant polynomial.

Since $P \not\equiv 0$, multiply by the denominators of the coefficients of P. We obtain another polynomial with integer coefficients, satisfying (1) and (2).

Moreover, if c is the leading coefficient of this last polynomial, then, using $(x \mapsto \frac{x}{c})$ and multiplying by c^{n-1} (where n is the degree of P), we obtain a monic polynomial, with (1) and (2).

Thus, with no loss of generality, we may suppose that $P \in \mathbb{Z}[x]$, where P has leading coefficient 1 (P is monic), and P satisfies (1) and (2).

The desired result follows immediately from this claim:

Claim: If $P \in \mathbb{Z}[x]$ is a monic polynomial, with degree greater than 1, then there exists an integer a such that $P(x) - a$ has a positive real irrational root.

Proof. Let p be a prime such that $p > P(1) - P(0)$ and greater than the largest real roots of $P(x) - P(0) - x$.

Let $a = p + P(0)$. Then $P(1) - a = P(1) - P(0) - p < 0$ and $P(p) - a = P(p) - P(0) - p > 0$.

From the Intermediate Value Theorem, it follows that $P(x) - a$ has a real root in $(1, p)$, say α.

Since $P(x) - a$ is a monic polynomial with integer coefficients, it is well known that, if α is a rational root of $P(x) - a$, then α divides $P(0) - a = -p$. Whence, $\alpha = 1$ or $\alpha = p$. Thus, α is irrational.

This ends the proof of the claim.

It follows from the claim that if $P \in \mathbb{Z}[x]$ is a monic polynomial satisfying (2), then P cannot have a degree greater than 1. That is, P is linear.

13th Iranian Mathematical Olympiad 1995

Does there exist a function $f : \mathbb{R} \to \mathbb{R}$ that fulfils all of the following conditions:

(a) $f(1) = 1$

(b) there exists $M > 0$ such that $-M < f(x) < M$

(c) if $x \neq 0$ then

$$f\left(x + \frac{1}{x^2}\right) = f(x) + \left(f\left(\frac{1}{x}\right)\right)^2 ?$$

Solution

Let n be the smallest integer for which $f(x) < n$ for all $x \neq 0$. Then, we can find $x \neq 0$ such that $f(x) \geq n - 1$. Then,

$$\left(f\left(\frac{1}{x}\right)\right)^2 = f\left(x + \frac{1}{x^2}\right) - f(x) < n - (n-1) = 1,$$

and thus, $f(\frac{1}{x}) > -1$. Now, substituting $\frac{1}{x}$ for x in the original equation, we have

$$(n-1)^2 \leq f(x)^2 = f\left(\frac{1}{x^2} + x\right) - f\left(\frac{1}{x}\right) < n + 1.$$

Thus, $(n-1)^2 < n + 1$, and thus, $n \in \{1, 2\}$. But, putting $x = 1$ in the original equation, we get $f(2) = 2$, and therefore, $n > 2$, a contradiction.

Estonian Mathematical Contest, 1995 (Final Round)

Prove that the polynomial $P_n(x) = 1 + x + \frac{x^2}{2} + \frac{x^3}{6} + \cdots + \frac{x^n}{n!}$ has no zeros if n is even and has exactly one zero if n is odd.

Solution

We shall prove by induction the following property

$$(\pi_n) \begin{cases} P_{2n} & \text{has no (real) zeros} \\ P_{2n+1} & \text{has exactly one (real) zero.} \end{cases}$$

Clearly (π_0) is true since $P_0(x) = 1$ and $P_1(x) = 1 + x$.

Suppose now that (π_n) is true for an integer $n \geq 0$. We will denote by x_n the unique zero of P_{2n+1}. The continuous function P_{2n} has no zeros and is positive for $x \geq 0$. Hence, $P_{2n}(x) > 0$ for all x.

Since the derivative P'_{2n+1} of P_{2n+1} is P_{2n}, the function P_{2n+1} is increasing. Furthermore, $\lim_{x \to -\infty} P_{2n+1}(x) = -\infty$ and $\lim_{x \to +\infty} P_{2n+1}(x) = +\infty$, so that $P_{2n+1}(x) < 0$ for $x < x_n$ and $P_{2n+1}(x) > 0$ for $x > x_n$.

Now, using $P'_{2n+2} = P_{2n+1}$, we see that $P_{2n+2}(x)$ is a minimum when $x = x_n$. Hence, for all x

$$\begin{aligned} P_{2n+2}(x) \geq P_{2n+2}(x_n) &= P_{2n+1}(x_n) + \frac{x_n^{2n+2}}{(2n+2)!} \\ &= \frac{x_n^{2n+2}}{(2n+2)!} = \frac{(x_n^{n+1})^2}{(2n+2)!} > 0. \end{aligned}$$

Thus, $P_{2n+2}(x)$ never takes the value 0.

Also, as above, P_{2n+3} is increasing, continuous, and $\lim_{x \to -\infty} P_{2n+3}(x) = -\infty$, $\lim_{x \to +\infty} P_{2n+3}(x) = +\infty$. Hence, P_{2n+2} has a unique zero. This completes the induction and shows that property (π_n) is true for all non-negative integer n.

Remark. As a supplement, we show that $\lim_{n \to \infty} x_n = -\infty$.

Consider

$$P_{2n+1}(-2n-3) = \sum_{k=0}^{2n+1} \frac{(-2n-3)^k}{k!} = \sum_{p=0}^{n} (2n+3)^{2p} \left(\frac{1}{(2p)!} - \frac{2n+3}{(2p+1)!} \right).$$

We have $P_{2n+1}(-2n-3) < 0$ (since $2n+3 > 2p+1$ for $p = 0, \ldots, n$) and thus, $x_n > -2n - 3$. Hence,

$$P_{2n+3}(x_n) = P_{2n+3}(x_n) - P_{2n+1}(x_n) = \frac{x_n^{2n+2}}{(2n+2)!}\left(1 + \frac{x_n}{2n+3}\right) > 0,$$

which implies $x_n > x_{n+1}$. Therefore, (x_n) is a decreasing sequence of real numbers and, as such, either $\lim_{n\to\infty} x_n = -\infty$ or (x_n) converges to a real number m. Assume that the latter does occur. Note that $m \leq x_n < 0$ for all n. Since for all $n \geq 0$ we have $P_{2n+1}(x) \leq e^x \leq P_{2n}(x)$ for all $x \leq 0$ [easy induction], we would have

$$\begin{aligned} 0 \leq e^{x_n} \leq P_{2n}(x_n) &= P_{2n+1}(x_n) - \frac{x_n^{2n+1}}{(2n+1)!} \\ &= -\frac{x_n^{2n+1}}{(2n+1)!} \leq -\frac{m^{2n+1}}{(2n+1)!}. \end{aligned}$$

But then, $\lim_{n\to\infty} e^{x_n} = 0$ (because $\lim\limits_{n\to\infty} \frac{m^{2n+1}}{(2n+1)!} = 0$) while, by the continuity of the exponential function, we must have $\lim\limits_{n\to\infty} e^{x_n} = e^m \neq 0$. This contradiction shows that $\lim\limits_{n\to\infty} x_n = -\infty$.

Estonian Mathematical Contest, 1995 (Final Round)

Find all functions $f : \mathbb{R} \to \mathbb{R}$ satisfying the following conditions for all $x \in \mathbb{R}$.
(a) $f(x) = -f(-x)$;
(b) $f(x+1) = f(x) + 1$;
(c) $f\left(\dfrac{1}{x}\right) = \dfrac{1}{x^2}f(x)$, if $x \neq 0$.

Solution

Set $g(x) = f(x) - x$. Then g satisfies
(a') $g(x) = -g(-x)$;
(b') $g(x+1) = g(x)$;
(c') $g\left(\dfrac{1}{x}\right) = \dfrac{1}{x^2}g(x)$, if $x \neq 0$.

By (a')–(b') we get $g(0) = g(-1) = 0$, and for all $x \neq 0, -1$ we have

$$\begin{aligned}
g(x) &= g(x+1) = (x+1)^2 g\left(\frac{1}{x+1}\right) = -(x+1)^2 g\left(-\frac{1}{x+1}\right) \\
&= -(x+1)^2 g\left(1 - \frac{1}{x+1}\right) = -(x+1)^2 g\left(\frac{x}{x+1}\right) \\
&= -(x+1)^2 \frac{x^2}{(x+1)^2} g\left(\frac{x+1}{x}\right) = -x^2 g\left(1 + \frac{1}{x}\right) = -x^2 g\left(\frac{1}{x}\right) \\
&= -g(x).
\end{aligned}$$

Hence, $g(x) \equiv 0$, and $f(x) = x$ for all $x \in \mathbb{R}$.

6th ROC Taiwan Mathematical Olympiad, 1997 (Part I)

Let a be a rational number, b, c, d be real, and the function $f : R \to [-1, 1]$ satisfying

$$f(x + a + b) - f(x + b) = c \cdot \lfloor x + 2a + \lfloor x \rfloor - 2\lfloor x + a \rfloor - \lfloor b \rfloor \rfloor + d$$

for each $x \in R$, where $\lfloor t \rfloor$ denotes the largest integer that is less than or equal to t. Show that f is a periodic function (that is, there is a positive number p such that $f(x + p) = f(x) \; \forall x \in R$).

Solution

If we replace x by $x - b + n$, where n is any integer, we obtain

$$\begin{aligned}
&f(x + n + a) - f(x + n) \\
&= c\lfloor x - b + n + 2a + \lfloor x - b + n \rfloor - 2\lfloor x - b + n + a \rfloor - \lfloor b \rfloor \rfloor + d \\
&= c\lfloor x - b + 2a + \lfloor x - b \rfloor - 2\lfloor x - b + a \rfloor - \lfloor b \rfloor \rfloor + d \\
&= f(x + a) - f(x) .
\end{aligned}$$

If $a = \frac{p}{q}$ with $q \neq 0$ and $(p, q) = 1$, then we claim that $f(x + aq) = f(x)$ for all $x \in \mathbb{R}$. Indeed, by the relation $f(x+n+a)-f(x+n) = f(x+a)-f(x)$ we have for all integers m

$$\begin{aligned}
f(x + maq) - f(x) &= m \sum_{i=1}^{q}(f(x + ai) - f(x + a(i - 1))) \\
&= m(f(x + aq) - f(x)) .
\end{aligned}$$

But, since $|f(x)| \leq 1$, we deduce that $f(x + aq) = f(x)$.

6th ROC Taiwan Mathematical Olympiad, 1997 (Part III)

Determine all the possible integers k such that there is a function $f : \mathbb{N} \longrightarrow \mathbb{Z}$ such that

(i) $f(1997) = 1998$,

(ii) $f(ab) = f(a) + f(b) + k \cdot f(d(a,b))$, $\forall a, b \in \mathbb{N}$, where $d(a,b)$ denotes the greatest common divisor of a and b.

Solution

Such a function exists if and only if $k = 0$ or $k = -1$.

First suppose that such an f exists. Then, for any prime number p, we have
$$f(p^2) = f(p) + f(p) + kf(p) = (k+2)f(p)$$
and
$$f(p^3) = f(p) + f(p^2) + kf(p) = (2k+3)f(p).$$
Hence, we may write $f(p^4) = f(p^2) + f(p^2) + kf(p^2) = (k+2)^2 f(p)$ as well as $f(p^4) = f(p) + f(p^3) + kf(p) = (3k+4)f(p)$.

Taking $p = 1997$, for which $f(p) \neq 0$, yields $(k+2)^2 = 3k+4$; hence $k = 0$ or $k = -1$.

Conversely, set $f(1) = 0$ and for $a > 1$ with standard factorization into primes $a = p_1^{s_1} \cdots p_r^{s_r}$ where $s_1, \ldots, s_r$ are positive integers and $p_1, \ldots, p_r$ are distinct prime numbers, set
$$f(a) = s_1(p_1 + 1) + \cdots + s_r(p_r + 1)$$
[respectively, $f(a) = (p_1 + 1) + \cdots + (p_r + 1) = p_1 + \cdots + p_r + r$].

Then, f satisfies conditions (i) (obviously) and (ii) with $k = 0$ [respectively, $k = -1$].

Indeed, (ii) is true when a or $b = 1$ and if $a = p_1^{s_1} \cdots p_r^{s_r}$, $b = q_1^{t_1} \cdots q_m^{t_m}$ are the standard factorization into primes of a and $b > 1$.

When a, b are coprime, we readily have $f(ab) = f(a) + f(b)$ in both cases, as desired (since $f(d(a,b)) = f(1) = 0$).

When $d(a,b) > 1$, we may suppose that $p_1 = q_1, \ldots, p_n = q_n$ ($n \leq \min(r,m)$) are the common prime factors of a and b. Then
$$ab = p_1^{s_1+t_1} \cdots p_n^{s_n+t_n} p_{n+1}^{s_{n+1}} \cdots p_r^{s_r} q_{n+1}^{t_{n+1}} \cdots q_m^{t_m}$$
so that
$$\begin{aligned} f(ab) &= (s_1 + t_1)(p_1 + 1) + \cdots + (s_n + t_n)(p_n + 1) + s_{n+1}(p_{n+1} + 1) \\ &\quad + \cdots + s_r(p_r + 1) + t_{n+1}(q_{n+1} + 1) + \cdots + t_m(q_m + 1) \\ &= f(a) + f(b) = f(a) + f(b) + 0 \cdot f(d(a,b)) \end{aligned}$$

[respectively,
$$\begin{aligned} f(ab) &= p_1 + \cdots + p_n + p_{n+1} + \cdots + p_r + q_{n+1} + \cdots + q_m + r + m - n \\ &= (p_1 + \cdots + p_r + r) + (q_1 + \cdots + q_m + m) - (q_1 + \cdots + q_n + n) \\ &= f(a) + f(b) - f(d(a,b))]. \end{aligned}$$

But, from $f(a^2) = (k+2)f(a)$, we also obtain $f(a^4) = (k+2)f(a^2) = (k+2)^2 f(a)$. Hence, $(k+2)^2 = 3k+4$ (we can take $a = 1997$), so that $f(a) \neq 0$, and such a function f exists for $k = 0$ and $k = -1$.

XXXIII Spanish Mathematical Olympiad, 1996

For each real number x, we denote by $\lfloor x \rfloor$ the biggest integer which is less than or equal to x. We define

$$q(n) = \left\lfloor \frac{n}{\lfloor \sqrt{n} \rfloor} \right\rfloor, \quad n = 1, 2, 3, \ldots.$$

(a) Forming a table with the values of $q(n)$ for $1 \leq n \leq 25$, make a conjecture about the numbers n for which $q(n) > q(n+1)$.

(b) Determine, with reasons, all the positive integer n such that

$$q(n) > q(n+1).$$

Solution

(a) The conjecture must be that $q(n) > q(n+1)$ if and only if $n = m^2 - 1$, where m is an integer $(m > 1)$.

(b) If $n = m^2 - 1$, then

$$q(n) = \left\lfloor \frac{m^2 - 1}{\lfloor \sqrt{m^2 - 1} \rfloor} \right\rfloor = \left\lfloor \frac{m^2 - 1}{m - 1} \right\rfloor = m + 1,$$

whereas

$$q(n+1) = \left\lfloor \frac{m^2}{\lfloor \sqrt{m^2} \rfloor} \right\rfloor = \left\lfloor \frac{m^2}{m} \right\rfloor = m < q(n).$$

Apart from these occasional decreases in the value of $q(n)$ when n is a perfect square, it is the case that $q(n+1) \geq q(n)$. To prove this, it is sufficient to show

$$q(m^2 + k) \geq q(m^2 + k - 1) \quad \text{for} \quad 1 \leq k \leq 2m.$$

This is in fact trivial, since $m^2 + k > m^2 + k - 1$ and $\lfloor \sqrt{m^2 + k} \rfloor = \lfloor \sqrt{m^2 + k - 1} \rfloor = m$ for such values of k.

20th Austrian-Polish Mathematical Competition, 1997

Let p_1, p_2, p_3, and p_4 be four distinct prime numbers. Prove that there does not exist a cubic polynomial $Q(x) = ax^3 + bx^2 + cx + d$ with integer coefficients such that

$$|Q(p_1)| = |Q(p_2)| = |Q(p_3)| = |Q(p_4)| = 3.$$

Solution

Let p_1, p_2, p_3, p_4 be four distinct prime numbers.

Suppose, for a contradiction, that there exists $Q(x) = ax^3 + bx^2 + cx + d$ with $a, b, c, d \in \mathbb{Z}$ and $a \neq 0$, such that $|Q(p_i)| = 3$ for $i = 1, 2, 3, 4$. With no loss of generality, we may suppose that at least two of the numbers $Q(p_i)$ are equal to 3 (otherwise we use $-Q$). Define $R(x) = Q(x) - 3$. Then $R \in \mathbb{Z}[x]$ and R has degree 3.

If $Q(p_i) = 3$ for each i: then the polynomial R has four distinct roots, which is impossible, since the degree of R is 3.

If exactly three of the $Q(p_i)$ are equal to 3: with no loss of generality, we may suppose that $Q(p_1) = Q(p_2) = Q(p_3) = 3$ and $Q(p_4) = -3$. Then p_1, p_2, p_3 are the roots of $R(n)$, and we have

$$R(x) = a(x - p_1)(x - p_2)(x - p_3).$$

Thus, $|R(p_4)| = |a|\,|p_4 - p_1|\,|p_4 - p_2|\,|p_4 - p_3| = 6 = 2 \times 3$, with $|a|$, $|p_4 - p_1|, |p_4 - p_2|, |p_4 - p_3| \in \mathbb{N}^*$. It follows that at least two of these four numbers are equal to 1.

But if $|p_i - p_j| = 1$, then the integers p_i and p_j are consecutive. And, if they are primes, they have to be 2 and 3. Since p_1, p_2, p_3, p_4 are distinct such a situation can occur at most one time. It follows that:

$|a| = 1$ and (by symmetry) $|p_4 - p_1| = 1$, $|p_4 - p_2| = 2$, $|p_4 - p_3| = 3$.

But the difference between two primes is odd if and only if one of these primes is 2. Then $p_4 = 2$. It follows that $p_2 \in \{0; 4\}$, which is impossible.

If exactly two of the $Q(p_i)$ are equal to 3, we may suppose that

$$Q(p_1) = Q(p_2) = 3 \quad \text{and} \quad Q(p_3) = Q(p_4) = -3.$$

Let α be the third root of $R(x)$. We then have $p_1 + p_2 + \alpha = -\frac{b}{a}$. Thus, $a\alpha$ is an integer.

Moreover:

$$|R(p_3)| = |p_3 - p_1|\,|p_3 - p_2|\,|ap_3 - a\alpha| = 6$$
$$|R(p_4)| = |p_4 - p_1|\,|p_4 - p_2|\,|ap_4 - a\alpha| = 6$$

with $|p_3 - p_1|, |p_3 - p_2|, |ap_3 - a\alpha|, |p_4 - p_1|, |p_4 - p_2|, |ap_4 - a\alpha| \in \mathbb{N}^*$.

As above, we deduce that:

(a) at least one of the integers $|p_4 - p_1|$ and $|p_4 - p_2|$ is equal to 1 or 3, and therefore, is odd.

(b) At least one of the integers $|p_3 - p_1|$ and $|p_3 - p_2|$ is equal to 1 or 3, and therefore, is odd.

Then:

- if $p_4 = 2$, we have $p_2 \neq 2$ and $p_1 \neq 2$. From (b), we must have $p_3 = 2 = p_4$. A contradiction.

- If $p_4 \neq 2$, then we may suppose that $p_1 = 2$ (from (a)). From (a) we deduce that $p_4 \in \{3, 5\}$, and from (b) that $p_3 \in \{3, 5\}$. We may suppose that $p_3 = 3$ and $p_4 = 5$. Then $p_2 \geq 7$ and $|p_2 - p_3| \geq 4$. Since $|p_2 - p_3|$ divides 6, we must have $p_2 - p_3 = 6$. Thus, $p_2 = 9$, which is not a prime, a contradiction.

And the proof is complete.

Chinese Mathematical Olympiad, 1997

Let $x_1, x_2, \ldots, x_{1997}$ be real numbers satisfying the following two conditions:

(a) $-\frac{1}{\sqrt{3}} \leq x_i \leq \sqrt{3}$ $(i = 1, 2, \ldots, 1997)$;

(b) $x_1 + x_2 + \cdots + x_{1997} = -318\sqrt{3}$.

Find the maximum of $x_1^{12} + x_2^{12} + \cdots + x_{1997}^{12}$ and give your reason.

Solution

Since the function $f(x) := x^{12}$ is convex, the maximum is attained when all the terms, perhaps except one, are equal to $\frac{-1}{\sqrt{3}}$ or $\sqrt{3}$. Suppose that n of the x_i's are equal to $\frac{-1}{\sqrt{3}}$, $1996 - n$ are equal to $\sqrt{3}$ and the last term is equal to

$$-318\sqrt{3} + \frac{n}{\sqrt{3}} - \sqrt{3}(1996 - n).$$

Since 1736 is the only integer which satisfies

$$\frac{-1}{\sqrt{3}} \leq -318\sqrt{3} + \frac{n}{\sqrt{3}} - \sqrt{3}(1996 - n) \leq \sqrt{3},$$

we deduce that the maximum is

$$\frac{1736^{12} + 260^{12} \times 3^{12} + 4^6}{3^6}.$$

Swedish Mathematical Competition, 1996 (Final Round)

For all integers $n \geq 1$ the functions p_n are defined for $x \geq 1$ by

$$p_n(x) = \frac{1}{2}\left(\left(x + \sqrt{x^2 - 1}\right)^n + \left(x - \sqrt{x^2 - 1}\right)^n\right).$$

Show that $p_n(x) \geq 1$ and that $p_{mn}(x) = p_m(p_n(x))$.

Solution

Let $m, n \in \mathbb{N}^*$, $x \geq 1$. Let $t = x + \sqrt{x^2 - 1}$. Then $t \geq 1$ and $\frac{1}{t} = x - \sqrt{x^2 - 1}$. It follows that

$$p_n(x) = \frac{1}{2}\left(t^n + \frac{1}{t^n}\right) \geq 1$$

since $\alpha + \frac{1}{\alpha} \geq 2$ for $\alpha > 0$. Note that equality occurs if and only if $t = 1$. That is, $x = 1$.

Moreover, we have

$$p_m(p_n(x)) = \frac{1}{2}\left(\left(\frac{1}{2}\left(t^n + \frac{1}{t^n}\right) + \sqrt{\frac{1}{4}\left(t^{2n} + \frac{1}{t^{2n}} + 2\right) - 1}\right)^m \right.$$
$$\left. + \left(\frac{1}{2}\left(t^n + \frac{1}{t^n}\right) - \sqrt{\frac{1}{4}(t^{2n} + \frac{1}{t^{2n}} + 2) - 1}\right)^m\right).$$

Since $\frac{1}{4}\left(t^{2n} + \frac{1}{t^{2n}} + 2\right) - 1 = \frac{1}{4}\left(t^n - \frac{1}{t^n}\right)^2$ with $t^n - \frac{1}{t^n} \geq 0$ (from $t \geq 1$), we deduce that

$p_m(p_n(x))$
$= \frac{1}{2}\left(\left(\frac{1}{2}t^n + \frac{1}{2t^n} + \frac{1}{2}t^n - \frac{1}{2t^n}\right)^m + \left(\frac{1}{2}t^n + \frac{1}{2t^n} - \frac{1}{2}t^n + \frac{1}{2t^n}\right)^m\right)$
$= \frac{1}{2}\left(t^{nm} + \frac{1}{t^{nm}}\right)$
$= \frac{1}{2}((x + \sqrt{x^2 - 1})^{nm} + (x - \sqrt{x^2 - 1})^{nm})$
$= p_{mn}(x)$ as desired.

Latvian Mathematical Olympiad, 1997 (1st TST)

Does there exist a function $f(x)$ which is defined for all reals and for which the identities

$$f(f(x)) = x \quad \text{and} \quad f(f(x)+1) = 1-x$$

hold?

Solution

We have that

$$f(f(x)) = x \tag{1}$$

and

$$f(f(x)+1) = 1-x \tag{2}$$

for every $x \in \mathbb{R}$. From (1) we conclude that f is a one-to-one function. Indeed: Let $a, b \in \mathbb{R}$ with $f(a) = f(b)$. Then

$$f(f(a)) = f(f(b)) \overset{(1)}{\Longleftrightarrow} a = b.$$

We put $x = 0$, $x = 1$ in (1) and (2) respectively, to obtain

$$f(f(0)) = 0 = f(f(1)+1) \tag{1-1}$$

$$f(0) = f(1)+1. \tag{3}$$

We put $x = 1$, $x = 0$ in (1) and (2) respectively. Thus, we have:

$$f(f(1)) = 1 = f(f(0)+1) \tag{1-2}$$

$$f(1) = f(0)+1 \tag{4}$$

Adding (3) and (4) we obtain: $0 = 2$ so that such a function does not exist.

Mathematical Olympiad in Bosnia and Herzegovina, 1997 (1st Day)

Let $f : A \to \mathbb{R}$, $A \subseteq \mathbb{R}$, be a function with the following characteristic:
$$f(x+y) = f(x) \cdot f(y) - f(xy) + 1, \quad (\forall x, y \in A).$$

(a) If $f : A \to \mathbb{R}$, $\mathbb{N} \subset A \subseteq \mathbb{R}$, is such a function, prove that the following is true:
$$f(n) = \begin{cases} \frac{c^{n+1}-1}{c-1}, & \forall n \in \mathbb{N}, \ c \neq 1, \\ n+1, & \forall n \in \mathbb{N}, \ c = 1, \end{cases}$$

$(c = f(1) - 1)$.

(b) Solve the given functional equation for $A = \mathbb{N}$.

(c) If $A = \mathbb{Q}$, find all the functions f which satisfy the given equation and the condition $f(1997) \neq f(1998)$.

Solution

(a) Let $A \subset \mathbb{R}$ such that $\mathbb{N} \subset A$ and $f : A \longrightarrow \mathbb{R}$ such that, for all $x, y \in A$,
$$f(x+y) = f(x)f(y) - f(xy) + 1. \tag{1}$$

Let $c = f(1) - 1$. Thus, $f(1) = c + 1$.

For $x = y = 0$, (1) leads to $f(0) = f^2(0) - f(0) + 1$ and then $f(0) = 1$.

For $y = 1$, (1) gives
$$f(x+1) = cf(x) + 1 \quad \text{for all } x \in A. \tag{2}$$

• If $c = 1$: From (2) we deduce $f(x+1) = f(x) + 1$ for all $x \in A$. And, from $f(0) = 1$, an easy induction leads to $f(n) = n+1$ for all $n \in \mathbb{N}$.

• If $c \neq 1$: Since $\frac{c^{n+2}-1}{c-1} = c\frac{c^{n+1}-1}{c-1} + 1$ and $f(0) = 1$, another easy induction leads to $f(x) = \frac{c^{n+1}-1}{c-1}$ for all $n \in \mathbb{N}$. Thus (a) is proved.

(b) We suppose now that $A = \mathbb{N}$. Let $f : A \longrightarrow \mathbb{R}$ which satisfies (1).

• If $c \neq 1$, using (a) and that $f(4) = f(2+2) = f^2(2) - f(4) + 1$, we must have
$$\frac{c^5-1}{c-1} = \left(\frac{c^3-1}{c-1}\right)^2 - \frac{c^5-1}{c-1} + 1$$

which is equivalent to $c^6 - 2c^5 + 2c^3 - c^2 = 0$. That is,
$$c^2(c-1)^3(c+1) = 0$$

and then $c \in \{0, -1\}$.

For $c = 0$ we have $f(n) = 1$ for all $n \in \mathbb{N}$.

For $c = -1$ we have $f(n) = 1$ if n is even, and $f(n) = 0$ if n is odd.

Conversely, it is easy to see that the three possibilities above give indeed three solutions. Then for $A = \mathbb{N}$ the functions which satisfy (1) are

(i) $f \equiv 1$

(ii) $f : (n \mapsto n+1)$

(iii) $f : (n \mapsto \frac{(-1)^n + 1}{2})$

(c) We now suppose that $A = \mathbb{Q}$. Let $f : A \longrightarrow \mathbb{R}$ which satisfies (1) and such that $f(1997) \neq f(1998)$.

Then the restriction of f to $\mathbb{N}$ satisfies (1) and from (b), must be of the form (i), (ii) or (iii). The condition $f(1997) \neq f(1998)$ eliminates (i).

In the two remaining cases we have $f(1) \in \{0, 2\}$ and $f(0) = 1$.

For $y = -x$, (1) leads to $f(0) = 1 = f(x)f(-x) - f(-x^2) + 1$. Thus,

$$f(x)f(-x) = f(-x^2) \quad \text{for all } x \in A. \tag{3}$$

In particular $f(1)f(-1) = f(-1)$. Since $f(1) \in \{0,2\}$, we have $f(-1) = 0$.

• If $f(1) = 0$ (case (iii)): Then, for all $x \in \mathbb{Q}$ we have:

$$f(1-x) = f(1)f(-x) - f(-x) + 1 = -f(-x) + 1$$

and

$$f(x-1) = f(-1)f(x) - f(-x) + 1 = -f(-x) + 1.$$

Thus, $f(x-1) = f(1-x)$.

Setting $X = x - 1$, we easily deduce that f is even. Then, from (3), we have $f^2(x) = f(x^2)$ for all $x \in \mathbb{Q}$. For $x = y$, (1) leads to

$$f(2x) = f^2(x) - f(x^2) + 1 = 1 \quad \text{for all } x \in \mathbb{Q}.$$

Then $f \equiv 1$. In particular $f(1997) = f(1998)$, which is a contradiction.

• If $f(1) = 2$ (case (ii)): From (2) we have $f(x+1) = f(x) + 1$ for all $x \in \mathbb{Q}$. By an easy induction, we deduce that:

$$f(x+p) = f(x) + p \quad \text{for all } x \in \mathbb{Q} \quad \text{and} \quad p \in \mathbb{Z}.$$

But

$$\begin{aligned} f(x+p) &= f(x)f(p) - f(px) + 1 \\ &= (p+1)f(x) - f(px) + 1 \quad \text{(from (b) (ii))}. \end{aligned}$$

Thus,

$$\begin{aligned} f(px) &= (p+1)f(x) + 1 - f(x) - p \\ &= pf(x) + 1 - p \quad \text{for all } x \in \mathbb{Q} \quad \text{and} \quad p \in \mathbb{Z}. \end{aligned}$$

Let a, b be integers with $b \neq 0$. From the above we then have $f(\frac{a}{b} \cdot b) = f(a) = a + 1$ and $f(\frac{a}{b} \cdot b) = bf(\frac{a}{b}) + 1 - b$. Then

$$f\left(\frac{a}{b}\right) = \frac{a}{b} + 1.$$

Thus, $f(x) = x + 1$ for all $x \in \mathbb{Q}$.

It is not difficult to verify that $f : (x \mapsto x + 1)$ is indeed a solution. Then, for $A = \mathbb{Q}$ the only function $f : A \longrightarrow \mathbb{R}$ which satisfies (1) and such that $f(1997) \neq f(1998)$ is $f : (x \mapsto x+1)$.

Fourth National Mathematical Olympiad of Turkey, 1997

Let $\mathbb{R}$ stand for the set of all real numbers. Show that there is no function $f : \mathbb{R} \to \mathbb{R}$ such that

$$f(x+y) > f(x)(1 + yf(x))$$

for all positive real x, y.

Solution

Suppose, for a contradiction, that such a function does exist. Then, for some positive x, y, we have

$$f(x+y) - f(x) > yf^2(x) \geq 0.$$

It follows that f is increasing on $\mathbb{R}^+$. Thus, f cannot be identically zero and there exists $\alpha > 0$ such that $f(\alpha) \neq 0$.

Since, for all $y > 0$, $f(x+y) > f(x) + yf^2(\alpha)$
with $\lim\limits_{y \to +\infty}(f(\alpha) + yf^2(\alpha)) = +\infty$, we deduce that $\lim\limits_{x \to +\infty} f(x) = +\infty$.
Then, we may consider $a > 0$ such that $f(a) > 0$. Thus, for $x \geq a$, we have $f(x) \geq f(a) > 0$. For such a real x, if $y = \frac{1}{f(x)}$, we have

$$f\left(x + \frac{1}{f(x)}\right) > 2f(x). \tag{1}$$

Let (x_n) be the sequence defined by $x_0 = a$ and $x_{n+1} = x_n + \frac{1}{f(x_n)}$ for $n \in \mathbb{N}$.

An easy induction shows that (x_n) is well defined and increasing. Thus, $x_n \geq a$ for all $n \geq 0$. For all $n \geq 0$, define $U_n = f(x_n)$. Therefore, $U_n > 0$ and, from (1), we have

$$\begin{aligned} U_{n+1} &= f(x_{n+1}) = f\left(x_n + \frac{1}{f(x_n)}\right) \\ &> 2f(x_n) = 2U_n. \end{aligned}$$

This leads to $U_n \geq 2^n U_0$ for all $n \geq 0$ and therefore,

$$\lim_{n \to +\infty} U_n = +\infty. \tag{2}$$

But, for all $n \geq 0$,

$$x_{n+1} - x_n = \frac{1}{f(x_n)} = \frac{1}{U_n} \leq \frac{1}{2^n U_0}.$$

Summing, we obtain:

$$x_{n+1} \leq a + \frac{1}{U_0}\sum_{k=0}^{n} \frac{1}{2^k} < a + \frac{1}{U_0}\sum_{k=0}^{+\infty} \frac{1}{2^k}.$$

Thus, $x_{n+1} < a + \frac{2}{U_0}$.

It follows that, for all $n \geq 0$,

$$U_{n+1} = f(x_{n+1}) < f\left(a + \frac{2}{U_0}\right),$$

where $f\left(a + \frac{2}{U_0}\right)$ is independent of n.

Thus, the sequence (U_n) is bounded from above, which contradicts (2). The conclusion now follows.

20th Austrian-Polish Mathematical Competition, 1997

Prove that there does not exist a function $f : \mathbb{Z} \to \mathbb{Z}$ such that $f(x + f(y)) = f(x) - y$ for all integers x and y.

Solution

Suppose that such a function exists. We denote by (R) the identity $f(x + f(y)) = f(x) - y$. Let $a = f(0)$. Suitable choices for x and y in (R) yield successively:

$$\begin{aligned} f(a) &= a \quad [x = y = 0] \\ f(2a) &= 0 \quad [x = y = a] \\ a &= 0 \quad [x = 0, y = 2a]; \text{ that is, } f(0) = 0. \end{aligned}$$

Taking $x = 0$ in (R), we obtain the identity

$$f(f(y)) = -y, \qquad (R')$$

valid for all integers y. From this, $-f(y) = f(f(f(y))) = f(-y)$, so that f is an odd function.

With the help of (R) and (R'), it follows that

$$f(u + v) = f(u + f(-f(v))) = f(u) + f(v)$$

for all integers u and v and, using induction, that $f(m) = mf(1)$ for all integers m. Taking $m = f(1)$, we get $f(f(1)) = (f(1))^2$ while (R') yields $f(f(1)) = -1$, a clear contradiction. The result follows.

Estonian Mathematical Olympiad, 1997 (Final Round)

A function f satisfies the condition
$$f(1) + f(2) + \cdots + f(n) = n^2 f(n)$$
for any positive integer n. Given that $f(1) = 999$, find $f(1997)$.

Solution

We have:
$$f(1) + \cdots + f(n) = n^2 f(n),$$
$$f(1) + \cdots + f(n-1) = (n-1)^2 f(n-1).$$
Subtracting and isolating $f(n)$, we get (for $n \geq 2$)
$$f(n) = \frac{(n-1)^2}{n^2 - 1} f(n-1) = \frac{n-1}{n+1} f(n-1).$$
By telescoping this formula, we obtain
$$f(n) = 2\frac{(n-1)!}{(n+1)!} f(1) = \frac{2f(1)}{(n+1)n}.$$
Since $f(1) = 999$, we have
$$f(1997) = \frac{2 \cdot 999}{(1998)(1997)} = \frac{1}{1997}.$$

Ukrainian Mathematical Olympiad, 1997

(10th Grade) Let $d(n)$ denote the greatest odd divisor of the natural number n. We define the function $f : \mathbb{N} \to \mathbb{N}$ as follows: $f(2n-1) = 2^n$, $f(2n) = n + \dfrac{2n}{d(n)}$ for all $n \in \mathbb{N}$.
Find all k such that $f(f(\ldots f(1)\ldots)) = 1997$, where f is iterated k times.

Solution

Let (x_k) be the sequence defined by $x_1 = 1$ and $x_{k+1} = f(x_k)$ for all $k \geq 1$. We want to find the integers k such that $x_{k+1} = 1997$.

The first terms of the sequence (x_k) are 1, 2, 3, 4, 6, 5, 8, 12, 10, 7, 16, 24, ... We present them in successive rows, $R_1, R_2, R_3, \ldots$, where R_j contains exactly j terms:

$R_1 : 1$
$R_2 : 2, 3$
$R_3 : 4, 6, 5$
$R_4 : 8, 12, 10, 7$
$R_5 : 16, 24, 20, 14, 9$
$\vdots$

We will prove that, for all positive integers i, j, with $j \leq i$, the jth number in R_i is $(2j-1)2^{i-j}$.

This is clearly true for $i = 1$. Let $i \geq 1$ be fixed. Suppose that the result is true for R_i. Then the last term of R_i (the one at the right) is $2i-1$. It follows that the first term of R_{i+1} is $f(2i-1) = 2^i = (2\times 1 - 1)2^{i+1-1}$. Thus, the desired formula is true for this first term.

Suppose that the result is true for the jth number in row R_{i+1}, where $1 \leq j < i+1$. Then the following term in R_{i+1} is:

$$f((2j-1)2^{i+1-j}) = (2j-1)2^{i-j} + 2^{i-j+1} = (2(j+1)-1)2^{i+1-(j+1)}.$$

Therefore, the formula is true for the value $j+1$. By induction, it is true for all $j \in \{1, 2, \ldots, i+1\}$.

Thus, the formula is true for all of R_{i+1}. By induction, it is true for all the rows.

Let $a = m2^n$, where m, n are non-negative integers and m is odd. Then $m2^n = (2j-1)2^{i-j}$ if and only if

$$\begin{cases} m = 2j-1 \\ n = i-j \end{cases} ; \quad \text{that is,} \quad \begin{cases} k = \dfrac{m+1}{2} \\ i = n + \dfrac{m+1}{2} \end{cases}.$$

It follows that a appears exactly once in the sequence, in position $\dfrac{m+1}{2}$ in $R_{n+\frac{m+1}{2}}$.

If $a = 1997$, then $m = 1997$ and $n = 0$. Thus, $k = i = 999$. Therefore, $x_k = 1997$ if and only if x_k is the last term of R_{999}, in which case

$$k = 1 + 2 + 3 + \cdots + 999 = 999 \cdot 500 = 499500.$$

Ukrainian Mathematical Olympiad, 1997

(11th Grade) Let $\mathbb{Q}^+$ denote the set of all positive rational numbers. Find all functions $f : \mathbb{Q}^+ \to \mathbb{Q}^+$ such that for all $x \in \mathbb{Q}^+$:

(a) $f(x+1) = f(x) + 1$,

(b) $f(x^2) = (f(x))^2$.

Solution

Let f be any such function. By property (a) and an immediate induction, we get
$$f(x+n) = f(x) + n \quad \text{for all } x \in \mathbb{Q}^+ \text{ and } n \in \mathbb{N}.$$
On the other hand,
$$(f(x+n))^2 = f((x+n)^2) = f(x^2 + 2nx + n^2)$$
$$= f(x^2 + 2nx) + n^2.$$
Comparing, we obtain
$$f(x^2 + 2nx) = f(x^2) + 2nf(x) \tag{1}$$
for all $x \in \mathbb{Q}^+$ and $n \in \mathbb{N}$.

Now, let $r = \frac{p}{q}$ be any element of $\mathbb{Q}^+$, where $p \in \mathbb{N}$ and $q \in \mathbb{N}$. From (1), with $x = r$ and $n = q$, we get $f(r^2 + 2p) = f(r^2) + 2qf(r)$. Then
$$f(r^2) + 2p = f(r^2) + 2qf(r),$$
which yields $f(r) = \frac{2p}{2q} = r$. Therefore, f is the identity on $\mathbb{Q}^+$.

Conversely, the identity of $\mathbb{Q}^+$ clearly satisfies conditions (a) and (b), whence it is the unique solution.

36th Armenian National Mathematical Olympiad, 1997

Let
$$p(x) = (x - a_1)^{n_1}(x - a_2)^{n_2}(x - a_3)^{n_3}$$
be a polynomial, such that
$$p(x) - 1 = (x - b_1)^{k_1}(x - b_2)^{k_2}(x - b_3)^{k_3},$$
where the numbers a_1, a_2, a_3, as well as b_1, b_2, b_3, are distinct, and n_1, n_2, n_3, k_1, k_2, k_3 are natural numbers. Prove that the degree of the polynomial $p(x)$ does not exceed 5.

Solution

First, note that since $p(a_i) = 0$ and $p(b_j) = 1$, we cannot have $a_i = b_j$ ($i, j = 1, 2, 3$). Thus, a_1, a_2, a_3, b_1, b_2, b_3 are six distinct numbers.

Now, if $(x-c)^m$ divides the polynomial $q(x)$, then $(x-c)^{m-1}$ divides its derivative $q'(x)$. Noticing that $(p(x) - 1)' = p'(x)$, we see that the polynomials $(x - a_i)^{n_i-1}$, $(x - b_j)^{k_j-1}$ divide $p'(x)$ (for $i, j = 1, 2, 3$). Since a_1, a_2, a_3, b_1, b_2, b_3 are six distinct numbers, the product
$$(x-a_1)^{n_1-1}(x-a_2)^{n_2-1}(x-a_3)^{n_3-1}(x-b_1)^{k_1-1}(x-b_2)^{k_2-1}(x-b_3)^{k_3-1}$$
divides $p'(x)$ as well. We deduce that
$$\begin{aligned}\deg p'(x) &\geq (n_1 - 1) + (n_2 - 1) + (n_3 - 1) \\ &\quad + (k_1 - 1) + (k_2 - 1) + (k_3 - 1) \\ &= 2\deg p(x) - 6,\end{aligned}$$
since $n_1+n_2+n_3 = k_1+k_2+k_3 = \deg p(x)$. Also, $\deg p'(x) = \deg p(x)-1$. The desired result $\deg p(x) \leq 5$ follows at once.

38th IMO Croation Team Selection Test, 1997

Let a, b, c, d be real numbers such that at least one is different from zero. Prove that all roots of the polynomial
$$P(x) = x^6 + ax^3 + bx^2 + cx + d$$
cannot be real.

Solution

More generally, let $a_0, a_1, \ldots, a_{n-3}$, $n \geq 3$, be real numbers such that at least one is different from zero. Then, the polynomial
$$P(x) = x^n + \sum_{j=0}^{n-3} a_j x^j \qquad (1)$$
cannot have only real roots. (The proposal's result is the particular case $n = 6$.)

Proof. Assume, by way of contradiction, that $P(x)$ has the (not necessarily distinct) real roots $x_1, x_2, \ldots, x_n$. Since $P(x)$ has leading coefficient 1,
$$P(x) = \prod_{k=1}^{n}(x - x_k). \qquad (2)$$
Comparing the coefficients of x^{n-1} and x^{n-2} gives
$$\sum_{k=1}^{n} x_k = 0 \quad \text{and} \quad \sum_{1 \leq j < k \leq n} x_j x_k = 0.$$
Hence,
$$\sum_{k=1}^{n} x_k^2 = \left(\sum_{k=1}^{n} x_k\right)^2 - 2 \sum_{1 \leq j < k \leq n} x_j x_k = 0,$$
which implies $x_1 = x_2 = \cdots = x_n = 0$, because $x_1, x_2, \ldots, x_n$ are all real. Thus, by (2), $P(x) = x^n$, and then, by (1), $a_0 = a_1 = \cdots = a_{n-3} = 0$, a contradiction.

Vietnamese Team Selection Test, 1997

4

Let $f : \mathbb{N} \to \mathbb{Z}$ be the function defined by:

$$f(0) = 2, \quad f(1) = 503,$$
$$f(n+2) = 503 f(n+1) - 1996 f(n) \quad \text{for all } n \in \mathbb{N}.$$

For every $k \in \mathbb{N}^*$, take k arbitrary integers $s_1, s_2, \ldots, s_k$ such that $s_i \geq k$ for all $i = 1, 2, \ldots, k$, and for every s_i ($i = 1, 2, \ldots, k$), take an arbitrary prime divisor $p(s_i)$ of $f(2^{s_i})$.

Prove that for positive integers $t \leq k$, we have:

$$\sum_{i=1}^{k} p(s_i) \mid 2^t \quad \text{if and only if} \quad k \mid 2^t.$$

Solution

The problem should be: prove that

$$2^t \,\Big|\, \sum_{i=1}^{k} p(s_i) \iff 2^t \mid k.$$

First, note that $f(n) = 4^n + 499^n$ for all $n \in \mathbb{N}$. (Use induction for example). Also note that if p is an odd prime, if $m, n \in \mathbb{N}$ are not divisible by p, and if $p \mid m^{2^s} + n^{2^s}$ ($s \geq 0$), then $p \equiv 1 \pmod{2^{s+1}}$. [*Editor's note:* Since p does not divide m, there is an integer a such that $am \equiv 1 \pmod p$. Since $m^{2^s} + n^{2^s} \equiv 0 \pmod p$, we have

$$(am)^{2^s} + (an)^{2^s} \equiv 0 \pmod p$$
$$(an)^{2^s} \equiv -1 \pmod p.$$

Therefore, $(an)^{2^{s+1}} \equiv 1 \pmod p$, which implies that the order of $an \pmod p$ is 2^{s+1}. By Fermat's Theorem, 2^{s+1} divides $p - 1$.]

Let $s_1, s_2, \ldots, s_k$ be k arbitrary integers such that $s_i \geq k$ for all $i = 1, 2, \ldots, k$, and let $p_i = p(s_i)$ be a prime divisor of $f(2^{s_i})$; that is, $p_i \mid 4^{2^{s_i}} + 499^{2^{s_i}}$ for each i. Therefore, $2^{s_i+1} \mid p_i - 1$, and in particular, $p_i \equiv 1 \pmod{2^k}$. Thus, $\sum_{i=1}^{k} p_i \equiv k \pmod{2^k}$. Consequently,

$$2^t \,\Big|\, \sum_{i=1}^{k} p(s_i) \iff 2^t \mid k.$$

Vietnamese Mathematical Olympiad, 1997

How many functions $f : \mathbb{N}^* \to \mathbb{N}^*$ are there that simultaneously satisfy the two following conditions:

(i) $f(1) = 1$,

(ii) $f(n) \cdot f(n+2) = (f(n+1))^2 + 1997$ for all $n \in \mathbb{N}^*$?

($\mathbb{N}^*$ denotes the set of all positive integers.)

Solution

Lemma. A sequence is defined by $a_1 = a$, $a_2 = b$, and
$$a_{n+2} = \frac{a_{n+1}^2 + c}{a_n}, \quad n = 1, 2, \ldots,$$
where a, b, c are real numbers and $c > 0$. Then all a_n ($n \geq 1$) are integers if and only if a, b, and $\dfrac{a^2 + b^2 + c}{ab}$ are integers.

Proof: If $a = 0$, then a_3 is not defined. Thus, $a \neq 0$. Similarly, if $b = 0$, then a_4 is not defined. Thus, $b \neq 0$. It follows inductively that $a_n \neq 0$, for all $n \geq 1$. More precisely, every term exists and is non-zero.

By the recurrence we find, for all $n \geq 2$,
$$a_{n+2}a_n - a_{n+1}^2 = c = a_{n+1}a_{n-1} - a_n^2,$$
and hence,
$$\frac{a_{n+2} + a_n}{a_{n+1}} = \frac{a_{n+1} + a_{n-1}}{a_n}.$$
Therefore, for all $n \geq 1$, we have
$$\frac{a_{n+2} + a_n}{a_{n+1}} = \frac{a_3 + a_1}{a_2} = \frac{\frac{b^2+c}{a} + a}{b} = \frac{a^2 + b^2 + c}{ab},$$
and hence,
$$a_{n+2} = \frac{a^2 + b^2 + c}{ab} \cdot a_{n+1} - a_n.$$
If a, b, $\dfrac{a^2 + b^2 + c}{ab}$ are integers, then an easy induction shows that a_n is an integer for all $n \geq 1$.

Conversely, assume that $a_n \in \mathbb{Z}$, for all $n \geq 1$. Then $a_1, a_2 \in \mathbb{Z}$ implies that $a, b \in \mathbb{Z}$. Moreover, we have $c \in \mathbb{Z}$, since $c = aa_3 - b^2$. Therefore, $\dfrac{a^2 + b^2 + c}{ab}$ is rational. Write $\dfrac{a^2 + b^2 + c}{ab} = \dfrac{p}{q}$, where $p \in \mathbb{Z}$, $q \in \mathbb{N}^*$ and $\gcd(p, q) = 1$. For $s \in \mathbb{N}^*$, we will prove inductively on s the following proposition $P(s)$: $q^s \mid a_n$, for all $n \geq s + 1$.

For all $k \geq 1$, the recurrence $a_{k+2} = \dfrac{p}{q} \cdot a_{k+1} - a_k$ gives $\dfrac{pa_{k+1}}{q} = a_{k+2} + a_k$, and hence, $q \mid (pa_{k+1})$. Since $\gcd(p, q) = 1$, it follows that $q \mid a_{k+1}$. Thus, $q \mid a_n$ for all $n \geq 2$, and $P(1)$ is true.

Suppose that $P(s)$ holds for some $s \geq 1$. Then $q^s \mid a_n$ for all $n \geq s+1$. Consider any $k \geq s+1$. Since $a_{k+2} = \frac{p}{q} a_{k+1} - a_k$, we have

$$\frac{a_{k+2} + a_k}{q^s} = \frac{p a_{k+1}}{q^{s+1}}.$$

By the induction hypothesis, $q^s \mid a_k$ and $q^s \mid a_{k+2}$; hence, $q^s \mid (a_{k+2} + a_k)$. Therefore, $q^{s+1} \mid (p a_{k+1})$. Since $\gcd(p, q^{s+1}) = 1$, we obtain $q^{s+1} \mid a_{k+1}$. Thus, $q^{s+1} \mid a_n$, for all $n \geq s+2$. This proves $P(s+1)$ and completes the induction.

Now let $s \geq 1$ be arbitrary. We have $c = a_{n+2} a_n - a_{n+1}^2$. For $n = s+1$, this yields $c = a_{s+3} a_{s+1} - a_{s+2}^2$. Using $P(s)$,

$$q^s \mid a_{s+1}, \quad q^s \mid a_{s+3}, \quad q^s \mid a_{s+2} \implies q^{2s} \mid (a_{s+3} a_{s+1} - a_{s+2}^2)$$
$$\implies q^{2s} \mid c.$$

Therefore,
$$q^{2s} \leq c \quad \text{for all } s \geq 1. \tag{1}$$

If we suppose $q > 1$, then $\lim_{s \to +\infty} q^{2s} = +\infty$, which contradicts (1). Thus, $q = 1$ and $\frac{a^2 + b^2 + c}{ab}$ is an integer. The lemma is proved.

We turn now to the initial problem.

Suppose $f : \mathbb{N}^* \longrightarrow \mathbb{N}^*$ is any function such that $f(1) = 1$ and

$$f(n+2) f(n) = (f(n+1))^2 + 1997, \quad \text{for all } n \in \mathbb{N}^*.$$

Let $b = f(2)$. Since $f(n)$ is an integer for all n, we have (from the lemma) that $b \in \mathbb{Z}$ and $\frac{1^2 + b^2 + 1997}{1 \cdot b} \in \mathbb{Z}$. Thus, $\frac{b^2 + 1998}{b} \in \mathbb{Z}$. Then $\frac{1998}{b} \in \mathbb{Z}$, and therefore, $b \mid 1998$. Thus, $f(2) = b$ is a positive divisor of 1998.

Conversely, let b be a positive divisor of 1998. Define $f : \mathbb{N}^* \longrightarrow \mathbb{R}$ by $f(1) = 1$, $f(2) = b$, and $f(n+2) \cdot f(n) = (f(n+1))^2 + 1997$. Since $f(1) \neq 0$ and $f(2) \neq 0$, each $f(n)$ exists and is non-zero (as in the proof of the lemma). Now $b \in \mathbb{Z}$, and

$$\frac{1^2 + b^2 + 1997}{1 \cdot b} = \frac{b^2 + 1998}{b} = b + \frac{1998}{b}$$

is an integer, since $b \mid 1998$. By the lemma, $f(n)$ is an integer for all $n \in \mathbb{N}^*$. Thus, $f : \mathbb{N}^* \longrightarrow \mathbb{Z}^*$. An easy induction shows that $f(n) > 0$ for every n, since $f(1) > 0$ and $f(2) > 0$. Thus, $f : \mathbb{N}^* \longrightarrow \mathbb{N}^*$, and we obtain an admissible sequence.

The above discussion reveals that the number of functions that satisfy both conditions (i) and (ii) in the problem is the same as the number of positive divisors of 1998. Since $1998 = 2 \cdot 3^3 \cdot 37$, the number of positive divisors of 1998 is $(1+1) \cdot (3+1) \cdot (1+1) = 16$. (It is known that the number of divisors of $p_1^{a_1} \cdots p_r^{a_r}$ is $(a_1 + 1) \cdots (a_r + 1)$.)

Vietnamese Mathematical Olympiad, 1997

(a) Find all polynomials of least degree, with rational coefficients, such that
$$f(\sqrt[3]{3}+\sqrt[3]{9}) = 3+\sqrt[3]{3}.$$

(b) Does there exist a polynomial with integer coefficients such that
$$f(\sqrt[3]{3}+\sqrt[3]{9}) = 3+\sqrt[3]{3}?$$

Solution

(a) Let $a = 3^{1/3}$.

Lemma. If $\alpha, \beta, \gamma \in \mathbb{Q}$ such that $\alpha a^2 + \beta a + \gamma = 0$, then $\alpha = \beta = \gamma = 0$.

Proof. Let $\alpha, \beta, \gamma \in \mathbb{Q}$ such that
$$\alpha a^2 + \beta a + \gamma = 0. \tag{1}$$
Suppose that $\alpha \neq 0$. Since a is a root of the quadratic $\alpha x^2 + \beta x + \gamma = 0$, we must have $a = \dfrac{-\beta \pm \sqrt{\Delta}}{2\alpha}$, where $\Delta = \beta^2 - 4\alpha\gamma \geq 0$. Note that $\sqrt{\Delta} \notin \mathbb{Q}$ (since $a \notin \mathbb{Q}$). Then
$$-24\alpha^3 = \beta^3 + 3\beta\Delta \pm \sqrt{\Delta}(3\beta^2 + \Delta).$$
It follows that $3\beta^2 + \Delta = 0$, which leads to $\beta = \Delta = 0$. Then we get $\alpha = 0$, a contradiction. Therefore, $\alpha = 0$, and equation (1) becomes $\beta a + \gamma = 0$. Since $a \notin \mathbb{Q}$, we deduce that $\beta = \gamma = 0$, and the lemma is proved.

Let $f \in \mathbb{Q}[x]$ such that $f(a + a^2) = a + 3$. If f has degree 1, then $f(x) = \alpha x + \beta$ for some $\alpha, \beta \in \mathbb{Q}$. Then, $\alpha(a + a^2) + \beta = a + 3$. Then the lemma implies that $\alpha = 0$ and $\alpha = 1$, which is clearly impossible. Therefore, f cannot have degree 1. If f has degree 2, then $f(x) = \alpha x^2 + \beta x + \gamma$ for some $\alpha, \beta, \gamma \in \mathbb{Q}$. Then, from the lemma, $f(a + a^2) = a + 3$ is equivalent to
$$\begin{aligned} \alpha + \beta &= 0 \\ 3\alpha + \beta &= 1 \\ 6\alpha \quad\quad + \gamma &= 3. \end{aligned}$$
The unique solution of this system is $\alpha = \frac{1}{2}$, $\beta = -\frac{1}{2}$, and $\gamma = 0$. It follows that there is a unique polynomial f of least degree having rational coefficients such that $f(\sqrt[3]{3}+\sqrt[3]{9}) = 3+\sqrt[3]{3}$, namely $f(x) = \frac{1}{2}x^2 - \frac{1}{2}x$.

(b) The answer is no.

Suppose, for the purpose of contradiction, that there exists $P \in \mathbb{Z}[x]$ such that $P(a + a^2) = a + 3$. Let $P(x) = \sum_{i=0}^{n} \alpha_i x^i$, where $\alpha_i \in \mathbb{Z}$ for each i. We must have $n \geq 3$, in view of our solution to (a). Note that $(a + a^2)^3 = 12 + 9(a + a^2)$. Then

$$P(a + a^2) = \sum_{i=0}^{2} \alpha_i (a + a^2)^i + (12 + 9(a + a^2)) \sum_{i=3}^{n} \alpha_i (a + a^2)^{i-3}.$$

It follows that the polynomial

$$Q(x) = \sum_{i=0}^{2} \alpha_i x^i + (12 + 9x) \sum_{i=3}^{n} \alpha_i x^{i-3}$$

satisfies $Q(a + a^2) = P(a + a^2) = a + 3$, where $Q \in \mathbb{Z}[x]$ with $\deg Q(x) = \deg P(x) - 2$. Now we can apply the same reasoning to Q in place of P. An easy induction leads to a polynomial R of degree at most 2, with integer coefficients, which satisfies $R(a + a^2) = a + 3$. From (a) we must have $R(x) = \frac{1}{2}x^2 - \frac{1}{2}x$, which does not have integer coefficients. This is a contradiction. The conclusion follows.

28th Austrian Mathematical Olympiad, 1997

Let n be a fixed natural number. Determine all polynomials $x^2 + ax + b$, where $a^2 \geq 4b$, such that $x^2 + ax + b$ divides $x^{2n} + ax^n + b$.

Solution

If $n = 1$, all polynomials $x^2 + ax + b$ are solutions. We will suppose $n > 1$ from now on. Since $a^2 \geq 4b$, there exist real numbers x_1, x_2 (not necessarily distinct) such that $x^2 + ax + b = (x - x_1)(x - x_2)$. It follows that $x^{2n} + ax^n + b = (x^n - x_1)(x^n - x_2)$.

Now, if $x^2 + ax + b$ divides $x^{2n} + ax^n + b$, then x_1, x_2 are roots of $x^{2n} + ax^n + b$, so that $x_1^n = x_1$ or $x_1^n = x_2$, and $x_2^n = x_2$ or $x_2^n = x_1$. Therefore, x_1 and x_2 must belong to $\{-1, 0, 1\}$.

Now we check the possible cases:

• If $x_1 = x_2 = 0$, then $a = b = 0$ and $x^2 + ax + b = x^2$ divides $x^{2n} + ax^n + b = x^{2n}$.

• If $x_1 = 0$, $x_2 = -1$, then $a = 1$, $b = 0$ and $x^2 + ax + b = x(x + 1)$ divides $x^n(x^n + 1)$ only if n is odd.

• If $x_1 = 0$, $x_2 = 1$, then $x^2 + ax + b = x(x - 1)$ divides $x^n(x^n - 1)$.

• If $x_1 = -1$, $x_2 = -1$, then $x^2 + ax + b = (x + 1)^2$ divides $(x^n + 1)^2$ only if n is odd.

• If $x_1 = 1$, $x_2 = 1$, then $x^2 + ax + b = (x - 1)^2$ divides $(x^n - 1)^2$.

• If $x_1 = -1$, $x_2 = 1$, then $x^2 + ax + b = x^2 - 1$ divides $x^{2n} - 1$.

In conclusion, for n odd $(n > 1)$, the solutions are x^2, $x(x+1)$, $x(x-1)$, $(x + 1)^2$, $(x - 1)^2$, $x^2 - 1$; and for n even, the solutions are x^2, $x(x - 1)$, $(x - 1)^2$, $x^2 - 1$.

Iranian Mathematical Olympiad, 1997 (Second Round)

4

Find all functions $f : \mathbb{N} \to \mathbb{N}\setminus\{1\}$ such that for all $n \in \mathbb{N}\setminus\{0\}$ we have,
$$f(n+1) + f(n+3) = f(n+5)f(n+7) - 1375.$$

Solution

Define $a_k = f(2k-1)$ and $b_k = f(2k)$. Then we have, for $k \in \mathbb{N}$,
$$a_k + a_{k+1} = a_{k+2}a_{k+3} - 1375, \qquad (1)$$
$$b_k + b_{k+1} = b_{k+2}b_{k+3} - 1375. \qquad (2)$$

Replacing k by $k+1$ in (1) and subtracting (1) from the resulting equation, we find that
$$a_{k+2} - a_k = a_{k+3}(a_{k+4} - a_{k+2}).$$
We know that $a_{k+3} \geq 2$. Therefore, $a_{k+2} = a_k$. Otherwise,
$$|a_{k+2} - a_k| > |a_{k+4} - a_{k+2}| > |a_{k+6} - a_{k+4}| > \cdots,$$
which is a contradiction. Taking $k = 1$ in (1) we obtain $a_1 + a_2 = a_1 a_2 - 1375$, or $(a_1 - 1)(a_2 - 1) = 1376$. Thus, $a_1 = t+1$ and $a_2 = \dfrac{1376}{t} + 1$, where t is any divisor of 1376. Therefore, the sequence satisfies the conditions if and only if
$$a_1 = a_3 = \cdots = t+1, \quad a_2 = a_4 = \cdots = \frac{1376}{t} + 1,$$
where t is any divisor of 1376.

Similarly, using (2),
$$b_1 = b_3 = \cdots = s+1, \quad b_2 = b_4 = \cdots = \frac{1376}{s} + 1,$$
where s is any divisor of 1376.

Finally, by combining the sequences, the function will be found.

Iranian Mathematical Olympiad, 1997 (Final Round)

Suppose that $f : \mathbb{R}^+ \to \mathbb{R}^+$ is a decreasing continuous function that fulfills the following condition for all $x, y \in \mathbb{R}^+$:

$$f(x+y) + f(f(x) + f(y)) = f\bigl(f(x+f(y)) + f(y+f(x))\bigr).$$

Prove that $f(x) = f^{-1}(x)$.

Solution

We first show that $\lim_{x \to +\infty} f(x) = 0$ and $\lim_{x \to 0} f(x) = +\infty$.

As f is decreasing and bounded below (by 0), f has a limit $l \geq 0$ as $x \to +\infty$. Suppose $l \geq 0$. From the given condition (GC) with $y = x$, we obtain

$$f(2x) + f(2f(x)) = f\bigl(2f(x+f(x))\bigr).$$

Letting $x \to +\infty$, the continuity of f yields $l + f(2l) = f(2l)$; whence, $l = 0$, a contradiction. Therefore, $l = 0$.

Similarly, when $x \to 0$, $f(x)$ tends to either $+\infty$ or a real number $m > 0$. Assume the latter, and consider $g(x) = f(x) - x$. Then g is decreasing and continuous, and $\lim_{x \to 0} g(x) = m > 0$, $\lim_{x \to +\infty} g(x) = -\infty$. Hence, the equation $g(x) = 0$ has a unique solution, say a, in $\mathbb{R}^+$. This means that a is the unique fixed point of f. Now, using GC,

$$f(x+a) + f(f(x)+a) = f\bigl(f(x+a) + f(a+f(x))\bigr).$$

The uniqueness of the fixed point implies that $f(x+a) + f(f(x)+a) = a$.

Letting $x \to 0$ in this relation, we get $f(a) + f(m+a) = a$, and hence, $f(m+a) = 0$. This is a contradiction, since the range of f is $\mathbb{R}^+$. Thus, $\lim_{x \to 0} f(x) = +\infty$.

It follows that the function f is continuous, decreasing from $(0, +\infty)$ onto $(0, +\infty)$ and, as such, is a bijection.

Lastly, letting $y \to 0$ in GC, we obtain, for all $x \in (0, +\infty)$,

$$f(x) + 0 = f\bigl(0 + f(f(x))\bigr), \quad \text{or} \quad f(x) = f\bigl(f(f(x))\bigr).$$

Since f is bijective, this yields $x = f(f(x))$, and the result $f(x) = f^{-1}(x)$ follows.

Swedish Mathematical Competition, 1997 (Final Round)

Let the sum of the two integers A and B be odd. Show that any integer can be written in the form $x^2 - y^2 + Ax + By$, where x and y are integers.

Solution

For integers x, y, let $f(x, y) = x^2 - y^2 + Ax + By$. If α is any integer, we have:
$$f(y + \alpha, y) = (A + B + 2\alpha)y + \alpha^2 + A\alpha.$$
Since $A + B$ is odd, we may choose $\alpha = \alpha_0$ such that $A + B + 2\alpha_0 = 1$. Let $k = \alpha_0^2 + A\alpha_0$. Note that k does not depend on y. Thus, $f(y + \alpha_0, y) = y + k$ takes every integer value when y ranges over $\mathbb{Z}$. The conclusion follows.

Swedish Mathematical Competition, 1997 (Final Round)

Let $s(m)$ denote the sum of the digits of the integer m. Prove that for any integer n, with $n > 1$ and $n \neq 10$, there is a unique integer $f(n) \geq 2$ such that $s(k) + s(f(n) - k) = n$ for all integers k satisfying $0 < k < f(n)$.

Solution

First suppose that $2 \leq n \leq 9$ and that $f(n) \geq 2$ satisfies the given condition. If $f(n) > n$, then we have $s(n) + s(f(n) - n) = n$, which is a contradiction, since $s(f(n) - n) > 0$ and $s(n) = n$. Thus, $f(n) \leq n$; whence, $2 \leq f(n) \leq 9$. From $s(1) + s(f(n) - 1) = n$, we now deduce that $1 + f(n) - 1 = n$, or $f(n) = n$. Conversely, $s(k) + s(n - k) = k + n - k = n$ if $0 < k < n \leq 9$.

If $n = 10$, it is readily checked that $f(n) = 10$ or $f(n) = 19$ are two possible solutions, so that the uniqueness fails in this case.

From now on, we assume $n \geq 11$. By the Euclidean Algorithm, we have $n = 9q + r$, where q is a positive integer and $r \in \{0, 1, 2, \ldots, 8\}$. We prove that the integer $f(n) = r \cdot 10^q + (10^q - 1)$ satisfies the condition of the problem.

Indeed, we can express $f(n)$ as

$$f(n) = r \cdot 10^q + 9 \cdot 10^{q-1} + \cdots + 9 \cdot 10 + 9,$$

and, if $0 < k < f(n)$, we can express k as

$$k = a_q \cdot 10^q + a_{q-1} \cdot 10^{q-1} + \cdots + a_1 \cdot 10 + a_0,$$

for some integers a_j such that $0 \leq a_q \leq r$ and $0 \leq a_j \leq 9$ for $j < q$. Hence,

$$f(n) - k = (r - a_q)10^q + (9 - a_{q-1})10^{q-1} + \cdots$$
$$\cdots + (9 - a_1)10 + (9 - a_0),$$

and thus, $s(f(n) - k) = r + 9q - (a_q + a_{q-1} + \cdots + a_0) = n - s(k)$, as desired.

It remains to show the uniqueness of $f(n)$. We first show that any suitable $f(n)$ must satisfy $f(n) \geq 11$. If we had $f(n) \leq 10$, then, using $s(1) + s(f(n) - 1) = n$, we would get $f(n) = n$, a contradiction, since $n \geq 11$ and $f(n) \leq 10$.

Next we show that $s(f(n)) = n$. From the given condition with $k = 9$ and with $k = 10$, we get $s(f(n) - 9) = n - 9$ and $s(f(n) - 10) = n - 1$. It follows that $n - 9 = s(f(n) - 10 + 1) = s(f(n) - 10) + 1 - 9m$, where m denotes the number of 9s at the right end of the integer $f(n) - 10$. (For example, $m = 1$ if $f(n) - 10 = 19129$). Hence,

$$n - 9 = (n - 1) + 1 - 9m = n - 9m.$$

Then $m = 1$. Thus, $f(n) - 10$ has a 9 as its rightmost digit and, consequently, so does $f(n)$. From this observation, we deduce that

$$s(f(n)) = s(f(n) - 1) + 1 = s(f(n) - 1) + s(1) = n.$$

If $f_1(n)$ and $f_2(n)$ were two suitable integers with $f_1(n) < f_2(n)$, then $s(f_1(n)) + s(f_2(n) - f_1(n)) = n$ yields $s(f_2(n) - f_1(n)) = 0$ (since $s(f_1(n)) = n$), a contradiction. The uniqueness of $f(n)$ follows.

Ukrainian Mathematical Olympiad, 1998 (11th Grade)

(11th grade) The function $f(x)$ is defined on $[0,1]$ and has values in $[0,1]$. It is known that $\lambda \in (0,1)$ exists such that $f(\lambda) \neq 0$ and $f(\lambda) \neq \lambda$. Also
$$f(f(x)+y) = f(x)+f(y)$$
for all x and y from the range of definition of the equality.

(a) Give an example of such a function.

(b) Prove that for any $x \in [0,1]$,
$$\underbrace{f(f(\ldots f(x)\ldots))}_{19} = \underbrace{f(f(\ldots f(x)\ldots))}_{98}.$$

Solution

(a) The function f defined by
$$f(x) = \begin{cases} 0 & \text{for } x=0 \text{ or } x \in (\tfrac{1}{2},1), \\ 1 & \text{otherwise}, \end{cases}$$
provides an example. Indeed, we can take $1/4$ for the required λ. Then $f(f(x)+y) = f(x)+f(y)$ is clearly satisfied when $f(x)=0$. If $f(x)=1$, we must have $y=0$ (since $f(x)+y$ must be in $[0,1]$); whence,
$$f(f(x)+y) = f(f(x)) = f(1) = 1 = f(x)+f(0) = f(x)+f(y).$$

(b) In the general case, let $f(0) = a \in [0,1]$. Then,
$$f(f(x)) = f(f(x)+0) = f(x)+a$$
for all $x \in [0,1]$. Taking $x=0$, we have $f(a) = 2a$. Consider now any positive integer n such that $na \leq 1$ and $f(na) = (n+1)a$. (We have just shown that one such integer is $n=1$.) Then $(n+1)a \leq 1$ (since f has values in $[0,1]$), and
$$f((n+1)a) = f(f(na)) = f(na)+a = (n+1)a+a = (n+2)a.$$
By induction, we have $na \leq 1$ for all positive integers n, which implies that $a=0$. Consequently, $f(f(x)) = f(x)$ for all $x \in [0,1]$. In other words, we have $f^2 = f$, which, by a simple induction, leads to $f^m = f$ for all positive integers m. (Here f^2 means $f \circ f$, etc.)

Thus, $f^{19} = f^{98}\ (= f)$, as required.

Vietnamese Mathematical Olympiad, 1998 (Category A, Day 1)

Let $a \geq 1$ be a real number. Define a sequence $\{x_n\}$ ($n = 1, 2, \ldots$) of real numbers by

$$x_1 = a, \quad x_{n+1} = 1 + \ln\left(\frac{x_n^2}{1 + \ln x_n}\right).$$

Prove that the sequence $\{x_n\}$ has a finite limit, and determine it.

Solution

We prove that the limit of the sequence $\{x_n\}$ is 1.

Let $f(x) = 1 + \ln\left(\frac{x^2}{1 + \ln x}\right)$, defined on $[1, \infty)$. We easily compute $f'(x) = \frac{1 + 2\ln x}{x(1 + \ln x)} > 0$ for $x \geq 1$. Thus, f is increasing on $[1, \infty)$.

Let $g(x) = f(x) - x$ on $[1, \infty)$. Then, we have $g'(x) = \frac{h(x)}{x(1 + \ln x)}$, where $h(x) = 1 + 2\ln x - x - x\ln x$. We have

$$h'(x) = -1 - \ln x - 1 + \frac{2}{x} = \frac{2 - 2x - x\ln x}{x} \leq 0,$$

with equality if and only if $x = 1$. Therefore, h is decreasing on $[1, \infty)$. Since $h(1) = 0$, we deduce that $h(x) \leq 0$ for $x \geq 1$, with equality if and only if $x = 1$.

It follows that $g'(x) \leq 0$ for $x \geq 1$, with equality if and only if $x = 1$. Then g is decreasing on $[1, \infty)$. Since $g(1) = 0$, we must have $g(x) \leq 0$ for $x \geq 1$, with equality if and only if $x = 1$. That is, $f(x) \leq x$, with equality if and only if $x = 1$.

Since $x_{n+1} = f(x_n)$ for $n \in \mathbb{N}^*$, an easy induction leads to the conclusion that the sequence $\{x_n\}$ is non-increasing and has a lower bound of 1. It follows that $\{x_n\}$ is convergent. Since f is continuous, the limit of $\{x_n\}$ is a fixed point of f. The unique fixed point is 1 (since $f(x) = x$ if and only if $x = 1$). Therefore, the limit is 1.

Vietnamese Mathematical Olympiad, 1998 (Category B, Day 1)

Prove that for each positive odd integer n there is exactly one polynomial $P(x)$ of degree n with real coefficients satisfying

$$P\left(x - \frac{1}{x}\right) = x^n - \frac{1}{x^n}$$

for all real $x \neq 0$.

Determine if the above assertion holds for positive even integers n.

Solution

The proof is by induction. The case $n = 1$ is trivial: simply choose $P(x) = P_1(x) = x$, and this polynomial is unique. Now suppose that, for $k = 1, 3, \ldots, 2n - 1$, we have a unique polynomial $P_k(x)$ of degree k with real coefficients such that

$$P_k\left(x - \frac{1}{x}\right) = x^k - \frac{1}{x^k}.$$

Using the Binomial Theorem, we have

$$\left(x - \frac{1}{x}\right)^{2n+1} = x^{2n+1} - \frac{1}{x^{2n+1}} - \binom{2n+1}{1}\left(x^{2n-1} - \frac{1}{x^{2n-1}}\right)$$
$$+ \cdots + (-1)^n \binom{2n+1}{n}\left(x - \frac{1}{x}\right).$$

It follows that

$$x^{2n+1} - \frac{1}{x^{2n+1}} = \left(x - \frac{1}{x}\right)^{2n+1} + \binom{2n+1}{1} P_{2n-1}\left(x - \frac{1}{x}\right)$$
$$- \cdots - (-1)^n \binom{2n+1}{n} P_1\left(x - \frac{1}{x}\right)$$
$$= P_{2n+1}\left(x - \frac{1}{x}\right).$$

We have constructed the unique polynomial explicitly by means of a recursive definition.

The assertion does not hold for even n. For $n = 2$ we would require

$$Q_2\left(x - \frac{1}{x}\right) = x^2 - \frac{1}{x^2},$$

and Q_2 would have to be of degree 2. Thus, we would require

$$a\left(x - \frac{1}{x}\right)^2 + b\left(x - \frac{1}{x}\right) + c = x^2 - \frac{1}{x^2}$$

for all $x \neq 0$. Setting $x = 1$ requires $c = 0$. Then, for $x \neq \pm 1$, we have

$$a\left(x - \frac{1}{x}\right) + b = x + \frac{1}{x}.$$

That is, $ax^2 - a + bx = x^2 + 1$. This would require $b = 0$, $a = 1$, and $a = -1$, which is an obvious contradiction.

1st Mediterranean Mathematical Olympiad, 1998

(a) Prove that the polynomial $z^{2n} + z^n + 1$, $n \in \mathbb{N}$, is divisible by the polynomial $z^2 + z + 1$ if and only if n is not a multiple of 3.

(b) Find the necessary and sufficient condition that the natural numbers p, q must satisfy for the polynomial $z^p + z^q + 1$ to be divisible by $z^2 + z + 1$.

Solution

We begin with the more general question (b) and show that the condition which we seek is $pq \equiv 2 \pmod{3}$.

The roots of $z^2 + z + 1$ are ω and ω^2, where $\omega = \exp(2\pi i/3)$. Hence, $z^p + z^q + 1$ is divisible by $z^2 + z + 1$ if and only if $\omega^p + \omega^q + 1 = 0$ and $\omega^{2p} + \omega^{2q} + 1 = 0$. Suppose that these conditions hold. Then

$$\begin{aligned} 0 = (\omega^p + \omega^q + 1)^2 &= \omega^{2p} + \omega^{2q} + 1 + 2(\omega^{p+q} + \omega^q + \omega^p) \\ &= 2(\omega^{p+q} - 1). \end{aligned}$$

Hence, $\omega^{p+q} = 1$, which implies that $p + q \equiv 0 \pmod{3}$. It follows that $\omega^q = \omega^{-p} = \omega^{2p}$. Then $\omega^p + \omega^{2p} + 1 = 0$, which implies that $p \equiv 1$ or $p \equiv 2 \pmod{3}$. Since $q \equiv -p \pmod{3}$, we get $pq \equiv 2 \pmod{3}$.

Conversely, if $pq \equiv 2 \pmod{3}$, say $p \equiv 1$ and $q \equiv 2$, then

$$\omega^p + \omega^q + 1 = \omega^{2p} + \omega^{2q} + 1 = \omega^2 + \omega + 1 = 0.$$

The conclusion follows.

As for (a), the result just obtained provides the condition $2n^2 \equiv 2$; that is, $n^2 \equiv 1 \pmod{3}$. This is clearly equivalent to $n \not\equiv 0 \pmod{3}$, which means n is not a multiple of 3.

Final National Selection Competition for Greek Team, 1998

4

(a) A polynomial $P(x)$ with integer coefficients takes the value -2 for seven distinct integer values of x. Prove that it cannot take the value 1996.

(b) Prove that there are irrational numbers x, y such that the number x^y is rational.

Solution

(a) Suppose $P(x)$ takes the value -2 for the distinct values x_1, x_2, x_3, x_4, x_5, x_6, x_7. Then, by the Remainder Theorem, we have

$$P(x) + 2 = (x - x_1)(x - x_2)(x - x_3) \cdots (x - x_7) Q(x),$$

where $Q(x)$ is also a polynomial with integer coefficients.

Suppose now that $P(k) = 1996$ for some integer k. Then

$$(k - x_1)(k - x_2) \cdots (k - x_7) Q(k) = 1998 = 2 \times 27 \times 37.$$

The factors $(k - x_i)$ are distinct, since the x_i are distinct. But the maximum number of distinct factors of 1998 whose product is 1998 is six. [*Ed.* To get the largest number of distinct factors, we could include 1, -1, 3, and -3; the remaining factor 3 would have to be combined with 2, 37, or 3 or -3 in order to maintain distinct factors, leaving us a maximum of six distinct factors. (For example, 1, -1, -3, 9, 2, 37, or 1, -1, 3, -3, 6, 37, etc.).] This contradiction shows that $P(k) \neq 1996$ for any integer k.

Hungary-Israel Mathematical Competition, 1999

Let $f(x)$ be a polynomial whose degree is at least 2. Define the sequence $g_i(x)$ by: $g_1(x) = f(x)$ and $g_{n+1}(x) = f(g_n(x))$ for $n = 1, 2, \ldots$. Let r_n be the average of the roots of $g_n(x)$. It is given that $r_{19} = 99$. Find r_{99}.

Solution

Let $f(x) = \sum_{i=0}^{p} c_i x^i$, where $p \geq 2$. Since $g_n = \underbrace{f \circ f \circ \ldots \circ f}_{n}$, we easily deduce that g_n is a polynomial of degree p^n. In $g_n(x)$, let α_n and β_n be the coefficients of x^{p^n} and x^{p^n-1}, respectively. Since g_n has p^n roots,

$$r_n = -\frac{\beta_n}{p^n \alpha_n}. \qquad (1)$$

For each $n \geq 1$,

$$g_{n+1}(x) = f(g_n(x)) = \sum_{i=0}^{p} c_i (g_n(x))^i = c_p (g_n(x))^p + \sum_{i=0}^{p-1} c_i (g_n(x))^i.$$

The degree of $\sum_{i=0}^{p-1} c_i (g_n(x))^i$ is at most $(p-1)p^n$. Therefore, terms of orders p^{n+1} and $p^{n+1} - 1$ appear only in the expansion of $c_p (g_n(x))^p$. We have

$$c_p (g_n(x))^p = c_p \left(\alpha_n x^{p^n} + \beta_n x^{p^n - 1} + h_n(x) \right)^p,$$

where $\deg(h_n) \leq p^n - 2$. Applying the Binomial Theorem, we deduce that

$$\alpha_{n+1} = c_p \alpha_n^p \quad \text{and} \quad \beta_{n+1} = p c_p \alpha_n^{p-1} \beta_n.$$

Then, using (1), we have

$$r_{n+1} = -\frac{\beta_{n+1}}{p^{n+1} \alpha_{n+1}} = -\frac{p c_p \alpha_n^{p-1} \beta_n}{p^{n+1} c_p \alpha_n^p} = -\frac{\beta_n}{p^n \alpha_n} = r_n.$$

By induction, $r_n = r_1$ for all $n \in \mathbb{N}$. Thus, $r_{99} = r_{19} = 99$.

Hungary-Israel Mathematical Competition, 1999

Find all the functions f from the set of rational numbers to the set of real numbers such that for all rational x, y,

$$f(x+y) = f(x)f(y) - f(xy) + 1.$$

Solution

The functions $x \mapsto 1$ and $x \mapsto x+1$ are clearly solutions. We now show that there is no other solution.

Suppose $f : \mathbb{Q} \to \mathbb{R}$ satisfies

$$f(x+y) = f(x)f(y) - f(xy) + 1 \tag{1}$$

for all x, $y \in \mathbb{Q}$. Taking $x = y = 0$, we get $f(0) = 1$. Then, taking $y = -x$ (for any $x \in \mathbb{Q}$), we get

$$f(x)f(-x) = f(-x^2). \tag{2}$$

From (2), $f(-1) = 0$ or $f(1) = 1$.

If $f(1) = 1$, then, using (1),

$$f(x) = f((x-1)+1) = f(x-1)f(1) - f(x-1) + 1 = 1,$$

for all $x \in \mathbb{Q}$. Hence, f is the constant function $x \mapsto 1$.

Suppose now that $f(-1) = 0$, and let $a = f(1)$. Taking $x = y = -1$ in (1) gives $f(-2) = (f(-1))^2 - f(1) + 1 = 1 - a$. Then, taking $x = 1$ and $y = -2$ in (1), we get

$$\begin{aligned} f(-1) &= f(1)f(-2) - f(-2) + 1, \\ 0 &= a(1-a) - (1-a) + 1, \\ &= a(2-a). \end{aligned}$$

Therefore, $a = 0$ or $a = 2$.

If $a = 0$ (that is, $f(1) = 0$), then, from (1),

$$\begin{aligned} f(x) &= f(x-1)f(1) - f(x-1) + 1 = 1 - f(x-1) \\ &= 1 - (f(x)f(-1) - f(-x) + 1) = f(-x), \end{aligned}$$

showing that f is even. Then (2) gives $(f(\tfrac{1}{2}))^2 = f(\tfrac{1}{4})$. It follows that

$$a = f\left(\tfrac{1}{2} + \tfrac{1}{2}\right) = \left(f\left(\tfrac{1}{2}\right)\right)^2 - f\left(\tfrac{1}{4}\right) + 1 = 1,$$

contradicting $a = 0$.

Thus, $f(1) = 2$. Using (1), we deduce that $f(x+1) = f(x) + 1$. By an easy induction, $f(x+n) = f(x) + n$ for all $x \in \mathbb{Q}$ and $n \in \mathbb{N}$. Recalling that $f(0) = 1$, we get $f(n) = n + 1$. Then, using (1) again,

$$\begin{aligned} f(x+n) &= f(x)f(n) - f(nx) + 1, \\ f(x) + n &= (n+1)f(x) - f(nx) + 1, \\ f(nx) &= nf(x) - n + 1. \end{aligned}$$

Now, let $r = m/n$, where $m, n \in \mathbb{N}$. Then
$$m + 1 = f(m) = f(nr) = nf(r) - n + 1.$$
Thus, $f(r) = \dfrac{m+n}{n} = 1 + r$. Moreover,
$$\begin{aligned} f(-r) &= f((1-r) + (-1)) = f(1-r)f(-1) - f(r-1) + 1 \\ &= -(f(r) - 1) + 1 = 2 - f(r) = 1 + (-r). \end{aligned}$$
As a result, $f(x) = x + 1$ for all $x \in \mathbb{Q}$, and the proof is complete.

Hungary-Israel Mathematical Competition, 1999

The function
$$f(x, y, z) = \frac{x^2 + y^2 + z^2}{x + y + z}$$
is defined for every x, y, z such that $x + y + z \neq 0$. Find a point (x_0, y_0, z_0) such that $0 < x_0^2 + y_0^2 + z_0^2 < \frac{1}{1999}$ and $1.999 < f(x_0, y_0, z_0) < 2$.

Solution

A solution is
$$(x_0, y_0, z_0) = (0.0009998, -0.0009998, 0.000001).$$

Note that $x_0 + y_0 + z_0 = 0.000001$ and $x_0^2 + y_0^2 + z_0^2 = 0.0000019992\ldots$ Thus, $f(x_0, y_0, z_0) = 1.9992\ldots$, which lies between 1.999 and 2.

(This solution required only one "difficult" calculation, namely $(0.0009998)^2 = (0.001)^2(1 - 0.0002)^2$ —which is not all that difficult.)

12th Korean Mathematical Olympiad, 1999

Suppose $f(x)$ is a function satisfying $|f(m+n) - f(m)| \leq \frac{n}{m}$ for all rational numbers n and m. Show that for all natural numbers k

$$\sum_{i=1}^{k} |f(2^k) - f(2^i)| \leq \frac{k(k-1)}{2}.$$

Solution

For all natural numbers i, we have

$$|f(2^{i+1}) - f(2^i)| = |f(2^i + 2^i) - f(2^i)| \leq \frac{2^i}{2^i} = 1.$$

For any integer $k > i$,

$$f(2^k) - f(2^i) = \sum_{j=i}^{k-1} \left(f(2^{j+1}) - f(2^j) \right),$$

and hence,

$$|f(2^k) - f(2^i)| \leq \sum_{j=i}^{k-1} |f(2^{j+1}) - f(2^j)| \leq \sum_{j=i}^{k-1} 1 = k - i.$$

Consequently,

$$\sum_{i=1}^{k} |f(2^k) - f(2^i)| = \sum_{i=1}^{k-1} |f(2^k) - f(2^i)| \leq \sum_{i=1}^{k-1} (k-i) = \frac{k(k-1)}{2}.$$

12th Korean Mathematical Olympiad, 1999

Suppose that for any real x ($|x| \neq 1$), a function $f(x)$ satisfies

$$f\left(\frac{x-3}{x+1}\right) + f\left(\frac{3+x}{1-x}\right) = x.$$

Find all possible $f(x)$.

Solution

Direct computation shows that the given condition is satisfied by

$$f(x) = \frac{7x + x^3}{2(1-x^2)}, \quad x \neq \pm 1.$$

We will prove that this is the only possible $f(x)$.

Let $f(x)$ be any function that satisfies the given condition. Let $x \in \mathbb{R}$ be arbitrary such that $x \neq \pm 1$. Set $y = \dfrac{x+3}{1-x}$. Then it is readily seen that $y \neq \pm 1$ and $x = \dfrac{y-3}{y+1}$. Furthermore, we find that $\dfrac{3+y}{1-y} = \dfrac{x-3}{x+1}$. Hence,

$$f(x) + f\left(\frac{x-3}{x+1}\right) = f\left(\frac{y-3}{y+1}\right) + f\left(\frac{3+y}{1-y}\right) = y;$$

that is,

$$f(x) + f\left(\frac{x-3}{x+1}\right) = \frac{x+3}{1-x}. \tag{1}$$

Now set $y = \dfrac{x-3}{x+1}$. It is readily seen that $y \neq \pm 1$ and $x = \dfrac{3+y}{1-y}$. Furthermore, $\dfrac{y-3}{y+1} = \dfrac{3+x}{1-x}$. Hence,

$$f\left(\frac{3+x}{1-x}\right) + f(x) = f\left(\frac{y-3}{y+1}\right) + f\left(\frac{3+y}{1-y}\right) = y;$$

that is,

$$f\left(\frac{3+x}{1-x}\right) + f(x) = \frac{x-3}{x+1}. \tag{2}$$

Adding (1) and (2), we then have

$$2f(x) + x = \frac{x+3}{1-x} + \frac{x-3}{x+1} = \frac{8x}{1-x^2},$$

from which it follows that

$$f(x) = \frac{1}{2}\left(\frac{8x}{1-x^2} - x\right) = \frac{7x + x^3}{2(1-x^2)}.$$

Grosman Memorial Mathematical Olympiad, 1999

Consider a polynomial $f(x) = x^4 + ax^3 + bx^2 + cx + d$ with integer coefficients a, b, c, d. Prove that if $f(x)$ has exactly one real root then $f(x)$ can be factored into terms with rational coefficients.

Solution

Since the coefficients are real, complex roots occur in conjugate pairs. Hence, if there is only one real root it must be either of multiplicity 2 or of multiplicity 4. If it is of multiplicity 4, then $f(x)$ must factor as $(x - k)^4$, where $a = 4k$, $b = 6k^2$, $c = 4k^3$, and $d = k^4$; thus, $k = a/4$ is rational (and must be an integer, since k^4 is an integer).

If x_0 is a real root of multiplicity 2, then $f(x_0) = 0$ and $f'(x_0) = 0$, and we have

$$x_0^4 + ax_0^3 + bx_0^2 + cx_0 + d = 0, \qquad (1)$$

$$4x_0^3 + 3ax_0^2 + 2bx_0 + c = 0. \qquad (2)$$

Subtracting x_0 times equation (2) from 4 times equation (1) gives

$$ax_0^3 + 2bx_0^2 + 3cx_0 + 4d = 0. \qquad (3)$$

Subtracting 4 times equation (3) from a times equation (2) gives

$$(3a^2 - 8b)x_0^2 + 2(ab - 6c)x_0 + (ac - 16d) = 0. \qquad (4)$$

Subtracting $4x_0$ times equation (4) from $(3a^2 - 8b)$ times equation (2) gives

$$(9a^3 - 32ab + 48c)x_0^2 + 2(3ba^2 - 8b^2 - 2ac + 32d)x_0, c(3a^2 - 8b) = 0. \qquad (5)$$

Now x_0^2 can be eliminated from (4) and (5) to deduce the value of x_0, which we perceive to be rational.

We now quote the result, which is well known, that if a, b, c, d are integers, then any rational root of the polynomial $x^4 + ax^3 + bx^2 + cx + d = 0$ is integral. We conclude that x_0 is an integer.

Now

$$x^4 + ax^3 + bx^2 + cx + d = (x - x_0)^2 q(x), \qquad (6)$$

where $q(x)$ is quadratic. This quadratic is irreducible over the reals, since the roots of $q(x) = 0$ are complex. Furthermore, from (6), since x_0 is integral, $q(x)$ is of the form $x^2 + Ax + B$, where A and B are rational. In fact, since $A - 2x_0 = a$ and $B = 2x_0 A + b - x_0^2$, we see that A and B are integers.

Russian Mathematical Olympiad, 1999 (11th Form)

A function $f : \mathbb{Q} \to \mathbb{Z}$ is considered. Prove that there exist two rational numbers a and b such that
$$\frac{f(a) + f(b)}{2} \leq f\left(\frac{a+b}{2}\right).$$

Solution

Assume the contrary (with the aim of reaching a contradiction). Then, for all distinct rational numbers a and b,
$$f\left(\frac{a+b}{2}\right) < \frac{f(a) + f(b)}{2}. \qquad (1)$$

This condition is not affected by adding a constant to the function f. Therefore, we may assume that $f(-1) \leq 0$ and $f(1) \leq 0$, without loss of generality.

We claim that for all $n \in \mathbb{N} \cup \{0\}$ and $x \in \{-2^{-n}, 0, 2^{-n}\}$, we have $f(x) \leq -n$. To prove this, we use induction on n. For $n = 0$, we must consider $x \in \{-1, 0, 1\}$. We have $f(\pm 1) \leq 0$, and hence, using (1),
$$f(0) = f\left(\frac{1-1}{2}\right) < \frac{f(1) + f(-1)}{2} \leq 0.$$

Thus, the claim is true for $n = 0$.

Now consider any fixed n for which the claim is true. Using (1) and the induction hypothesis, we have
$$f(\pm 2^{-n-1}) = f\left(\frac{0 \pm 2^{-n}}{2}\right) < \frac{f(0) + f(\pm 2^{-n})}{2} \leq \frac{-n-n}{2} = -n.$$

Since f takes only integer values, we must have $f(\pm 2^{-n-1}) \leq -(n+1)$. Then, using (1) again,
$$f(0) = f\left(\frac{2^{-n-1} - 2^{-n-1}}{2}\right) < \frac{f(2^{-n-1}) + f(-2^{-n-1})}{2} \leq -(n+1).$$

This completes the induction and proves the claim.

We have shown that $f(0) \leq -n$ for all non-negative integers n. This is impossible. Therefore, there is no function $f : \mathbb{Q} \to \mathbb{Z}$ that satisfies (1) for all distinct rational numbers a and b.

16th Iranian Mathematical Olympiad, 1999 (Second Round)

Find all functions $f : \mathbb{R} \to \mathbb{R}$ satisfying,

$$f(f(x) + y) = f(x^2 - y) + 4f(x)y,$$

for all real numbers $x, y \in \mathbb{R}$.

Solution

The functions $x \mapsto 0$ and $x \mapsto x^2$ are clearly solutions. We now show that there is no other solution.

Suppose $f : \mathbb{R} \to \mathbb{R}$ satisfies

$$f(f(x) + y) = f(x^2 - y) + 4f(x)y \qquad (1)$$

for all $x, y \in \mathbb{R}$. Let $a = f(0)$.

Taking $x = 0$ in (1) gives

$$f(a + y) = f(-y) + 4ay \qquad (2)$$

for all y. In (2), we first take $y = 0$ to get $f(a) = a$, then $y = -a$ to get $a = a - 4a^2$. It follows that $a = 0$. Then, from (2), f is an even function. Comparing the results of the substitutions $y = -f(x)$ and $y = x^2$ in (1) easily leads to $4\bigl(f(x)\bigr)^2 = 4x^2 f(x)$. Thus, $f(x) = 0$ or $f(x) = x^2$.

Assume now that there exists x_0 such that $f(x_0) \neq 0$. Then $x_0 \neq 0$ and $f(x_0) = x_0^2$. Since f is even, we may suppose that $x_0 > 0$. Let x be any non-zero real number. By (1) with $y = -x_0$, we obtain

$$f(f(x) - x_0) = f(x^2 + x_0) - 4f(x)x_0.$$

If $f(x) = 0$, then

$$f(x^2 + x_0) = f(-x_0) = f(x_0) = x_0^2 \neq 0.$$

This implies that $f(x^2 + x_0) = (x^2 + x_0)^2$, and thus, $(x^2 + x_0)^2 = x_0^2$. This is not possible, since $x_0 > 0$ and $x \neq 0$. Therefore, $f(x) = x^2$.

Thus, $f(x) = 0$ for all x or $f(x) = x^2$ for all x.

Vietnamese Mathematical Olympiad, 1999 (Category B)

Let $f(x)$ be a continuous function defined on $[0, 1]$ such that
(i) $f(0) = f(1) = 0$,
(ii) $2f(x) + f(y) = 3f\left(\frac{2x+y}{3}\right)$ $\forall\, x, y \in [0, 1]$.

Prove that $f(x) = 0$ for all $x \in [0, 1]$.

Solution

Since $|f|$ is continuous on the closed and bounded interval $[0, 1]$, there exists a real number $a \in [0, 1]$ at which $|f|$ attains its maximum M; that is, $|f(x)| \leq |f(a)| = M$ for all $x \in [0, 1]$. We will show that $M = 0$, which implies that f is the zero function.

Case 1: $0 \leq a \leq \frac{1}{2}$.

Let $a_1 = 2a/3$ and $b_1 = 5a/3$. Then $0 \leq a_1 \leq a \leq b_1 < 1$ and $a = \frac{2a_1 + b_1}{3}$. Hence, using (ii),

$$\begin{aligned} M = |f(a)| &= \left|f\left(\frac{2a_1 + b_1}{3}\right)\right| = \left|\tfrac{2}{3}f(a_1) + \tfrac{1}{3}f(b_1)\right| \\ &\leq \tfrac{2}{3}|f(a_1)| + \tfrac{1}{3}|f(b_1)| \leq \tfrac{2}{3}M + \tfrac{1}{3}M = M, \end{aligned}$$

from which we deduce that $|f(a_1)| = M$.

Iterating, we construct a sequence $\{a_n\}_{n=1}^{\infty}$ such that $a_n = (\tfrac{2}{3})^n a$ and $|f(a_n)| = M$ for all positive integers n. Since $\{a_n\}$ converges to 0, the continuity of f implies that $M = \lim_{n \to \infty} |f(a_n)| = |f(0)| = 0$.

Case 2: $\frac{1}{2} < a \leq 1$.

Let $g(x) = f(1-x)$. Then g satisfies all the given conditions on f, and the maximum of $|g|$ is M, attained at $1 - a$. Since $0 \leq 1 - a < \frac{1}{2}$, we may apply Case 1 to the function g to deduce that $M = 0$.

Turkish Team Selection Test for 40th IMO, 1999

Determine all functions $f : \mathbb{R} \longrightarrow \mathbb{R}$ such that the set
$$\left\{ \frac{f(x)}{x} : x \neq 0 \text{ and } x \in \mathbb{R} \right\}$$
is finite, and for all $x \in \mathbb{R}$
$$f(x - 1 - f(x)) = f(x) - x - 1.$$

Solution

We shall show that f must be the identity function on $\mathbb{R}$. Note that the identity satisfies all the given conditions.

Let $f : \mathbb{R} \longrightarrow \mathbb{R}$ satisfy the given conditions, and define the set
$$Y = \{y \mid y = x - f(x), \ x \in \mathbb{R}\}.$$
If $y \in Y$, then $y = x - f(x)$ for some $x \in \mathbb{R}$, and hence,
$$f(y - 1) = f(x - f(x) - 1) = f(x) - x - 1 = y - 1. \qquad (1)$$
Then $2y = (y - 1) - f(y - 1)$, and therefore $2y \in Y$.

Consider any fixed $a \in Y$. An easy induction implies that $2^n a \in Y$ for all $n = 0, 1, 2, \ldots$. From the given conditions on f, we know that the following set is finite:
$$\left\{ \frac{f(2^n a - 1)}{2^n a - 1} : n \in \mathbb{N}, a \neq 2^{-n} \right\}.$$
Therefore, we may choose two positive integers m and n, with $m \neq n$, $a \neq 2^{-m}$, and $a \neq 2^{-n}$ such that
$$\frac{f(2^n a - 1)}{2^n a - 1} = \frac{f(2^m a - 1)}{2^m a - 1}.$$
Using (1), we have
$$\frac{-2^n a - 1}{2^n a - 1} = \frac{-2^m a - 1}{2^m a - 1},$$
which forces $2^n a = 2^m a$, and hence $a = 0$. Then $Y = \{0\}$, proving that f is the identity function.

Japanese Mathematical Olympiad, 1999 (Final Round)

Let $f(x) = x^3 + 17$. Prove that for each natural number n, $n \geq 2$, there is a natural number x, for which $f(x)$ is divisible by 3^n but not by 3^{n+1}.

Solution

First we prove two lemmas.

Lemma 1. Let $n \geq 2$. Let $x \in \mathbb{N}$ such that $x^3 + 17 \equiv 0 \pmod{3^n}$ and $x^3 + 17 \not\equiv 0 \pmod{3^{n+1}}$. Then either

$$(x + 3^{n-1})^3 + 17 \equiv 0 \pmod{3^{n+1}}$$
$$\text{or} \quad (x + 2 \cdot 3^{n-1})^3 + 17 \equiv 0 \pmod{3^{n+1}}.$$

Proof: Since $3 \mid (x^3 + 17)$, we see that $x \equiv 1 \pmod{3}$. Then $x^2 = 3m + 1$ for some $m \in \mathbb{N}$. Also, $n \geq 2$ gives $2n - 1 \geq n + 1$ and $3n - 3 \geq n + 1$. By assumption, we have $x^3 + 17 = 3^n k$ and $k = 3q + 1$ or $k = 3q + 2$.

If $k = 3q + 1$, then

$$\begin{aligned}
(x + 2 \cdot 3^{n-1})^3 + 17 &= x^3 + 17 + 2 \cdot x^2 \cdot 3^n + 4 \cdot x \cdot 3^{2n-1} + 8 \cdot 3^{3n-3} \\
&= 3^n(3q + 1 + 6m + 2 + 3y) \\
&= 3^{n+1}(q + 2m + y + 1) \\
&\equiv 0 \pmod{3^{n+1}}.
\end{aligned}$$

If $k = 3q + 2$, then

$$\begin{aligned}
(x + 3^{n-1})^3 + 17 &= x^3 + 17 + x^2 \cdot 3^n + x \cdot 3^{2n-1} + 3^{3n-3} \\
&= 3^n(3q + 2 + 3m + 1 + 3y) \\
&= 3^{n+1}(q + m + y + 1) \\
&\equiv 0 \pmod{3^{n+1}}.
\end{aligned}$$

∎

Lemma 2. Let $n \geq 2$. Let $x \in \mathbb{N}$ such that $x^3 + 17 \equiv 0 \pmod{3^{n+1}}$. Then

$$(x + 3^{n-1})^3 + 17 \equiv 0 \pmod{3^n}$$
$$\text{and} \quad (x + 3^{n-1})^3 + 17 \not\equiv 0 \pmod{3^{n+1}}.$$

Proof: As in the proof of Lemma 1, we have $x^2 = 3m + 1$ for some $m \in \mathbb{N}$. Since $n \geq 2$, we have $2n - 1 \geq n + 1$ and $3n - 3 \geq n + 1$. By assumption, $x^3 + 17 = 3^{n+1} \cdot k$. Hence,

$$\begin{aligned}
(x + 3^{n-1})^3 + 17 &= 3^{n+1} \cdot k + 3^n \cdot x^2 + 3^{2n-1} \cdot x + 3^{3n-3} \\
&= 3^n(3k + 3m + 1 + 3y),
\end{aligned}$$

from which the desired result follows. ∎

Now we solve the given problem by constructing a sequence of natural numbers x_n such that $3^n \mid (x_n^3 + 17)$ and $3^{n+1} \nmid (x_n^3 + 17)$, for all n. We proceed recursively, starting the sequence with $x_2 = 1$. Note that

$$x_2^3 + 17 = 18 \equiv 0 \pmod{3^2} \quad \text{and} \quad x_2^3 + 17 \not\equiv 0 \pmod{3^3}.$$

Suppose x_n has been defined such that $3^n \mid (x_n^3 + 17)$ and $3^{n+1} \nmid (x_n^3 + 17)$. By Lemma 1, we can let $\tilde{x}_{n+1} = x_n + 3^{n-1}$ or $\tilde{x}_{n+1} = x_n + 2 \cdot 3^{n-1}$ to obtain $3^{n+1} \mid (\tilde{x}_{n+1}^3 + 17)$. Then we define

$$x_{n+1} = \begin{cases} \tilde{x}_{n+1} & \text{if } 3^{n+2} \nmid (\tilde{x}_{n+1}^3 + 17), \\ \tilde{x}_{n+1} + 3^n & \text{if } 3^{n+2} \mid (\tilde{x}_{n+1}^3 + 17). \end{cases}$$

By Lemma 2, we see that $3^{n+1} \mid (x_{n+1}^3 + 17)$ but $3^{n+2} \nmid (x_{n+1}^3 + 17)$.

Japanese Mathematical Olympiad, 1999 (Final Round)

Prove that
$$f(x) = (x^2 + 1^2)(x^2 + 2^2)(x^2 + 3^2) \cdots (x^2 + n^2) + 1$$
cannot be expressed as a product of two integral-coefficient polynomials with degree greater than 1.

Solution

For the purpose of contradiction, suppose that $f(x) = P(x) \cdot Q(x)$ where $P(x)$ and $Q(x)$ are integral-coefficient polynomials whose degrees are greater than 1.

The complex number ki satisfies the equation $P(ki)Q(ki) = 1$ for $k = \pm 1, \pm 2, \ldots, \pm n$. Since the complex numbers $P(ki)$, $Q(ki)$ are of the form $a + bi$ with $a, b \in \mathbb{Z}$, we must have

$$(P(ki), Q(ki)) \in \{(1,1), (-1,-1), (i,-i), (-i,i)\}. \tag{1}$$

In all cases, $P(ki) = \overline{Q(ki)} = Q(-ki)$. Thus, the polynomial $P(x) - Q(-x)$ has at least $2n$ distinct roots (the complex numbers $\pm i, \pm 2i, \ldots, \pm ni$), while its degree is less than $2n$. Therefore, $P(x) - Q(-x)$ is the zero polynomial; that is, $P(x) = Q(-x)$. Hence,

$$\deg P(x) = \deg Q(x) = n.$$

Since $f(x)$ is monic (that is, having 1 as the coefficient of the highest power of x), we may suppose that $P(x)$ and $Q(x)$ are both monic. Then the polynomial $(P(x))^2 - (Q(x))^2$ has degree less than $2n$. This polynomial has at least the $2n$ distinct roots ki (for $k = \pm 1, \pm 2, \ldots, \pm n$), because of (1). Hence, $(P(x))^2 - (Q(x))^2 = 0$. We cannot have $P(x) = -Q(x)$, since $P(x)$ and $Q(x)$ are both monic. We conclude that $P(x) = Q(x)$. Then $f(x) = (P(x))^2$. This implies that $(P(0))^2 = f(0) = (n!)^2 + 1$, which is impossible with $P(0) \in \mathbb{Z}$ and $n \geq 1$. This contradiction establishes the result.

Swiss Mathematical Contest, 1999 (First Day)

Determine all functions $f : \mathbb{R} \setminus \{0\} \to \mathbb{R}$, satisfying
$$\frac{1}{x}f(-x) + f\left(\frac{1}{x}\right) = x \text{ for all } x \in \mathbb{R} \setminus \{0\}.$$

Solution

Let f be a solution. Replacing x by $-1/x$ in the given equation, we get
$$-xf\left(\frac{1}{x}\right) + f(-x) = -\frac{1}{x},$$
and hence
$$-f\left(\frac{1}{x}\right) + \frac{1}{x}f(-x) = -\frac{1}{x^2}.$$
Adding this equation to the given equation gives $\frac{2}{x}f(-x) = x - \frac{1}{x^2}$. Then $f(-x) = \frac{x^2}{2} - \frac{1}{2x}$, and finally, $f(x) = \frac{x^2}{2} + \frac{1}{2x}$ for all $x \in \mathbb{R}\setminus\{0\}$.

Conversely, it is easy to check that the function $x \mapsto f(x) = \frac{x^2}{2} + \frac{1}{2x}$ for all $x \in \mathbb{R}\setminus\{0\}$, is actually a solution. Thus, it is the unique solution to the problem.

Swiss Mathematical Contest, 1999 (Second Day)

Prove that for every polynomial $P(x)$ of degree 10 with integer coefficients there is an (in both directions) infinite arithmetic progression which does not contain $P(k)$ for any integer k.

Solution

More generally, we claim that the result holds for any polynomial $P(x)$ with integer coefficients which is *not* of the form $P(x) = mx + p$ for some integers m and p with $m \in \{-1, 1\}$.

Lemma. Let $P(x) = \sum_{i=0}^{q} a_i x^i$, with $q \geq 1$, $a_i \in \mathbb{Z}$ for all i, and $a_q > 0$. Then there exists an integer n such that $P(x+n) = \sum_{i=0}^{q} b_i x^i$, with $b_i \in \mathbb{Z}$ for all i, $b_q = a_q$, and $b_i \geq 3$ for all $i < q$.

Proof of Lemma. If $a_i \geq 3$ for all $i < q$, then we just choose $n = 0$ and we are done. Therefore, we can assume that there is some $i < q$ such that $a_i < 3$. Let t be the greatest such i. Then $a_i \geq 3$ for $t < i < q$.

Let $n \in \mathbb{Z}$. The polynomial $P(x+n)$ clearly has integer coefficients and degree equal to q. Hence, $P(x+n) = \sum_{i=0}^{q} b_i x^i$, where $b_i \in \mathbb{Z}$ for all i. Furthermore, $b_q = a_q$, since x^q appears only in the term $a_q(x+n)^q$.

Now consider any fixed integer j such that $t < j < q$. In $P(x+n)$, the power x^j appears in the terms $a_i(x+n)^i$ only for $i \geq j$. Since $a_i > 0$ for $i > t$, we have $b_j \geq a_j \geq 3$.

By similar reasoning, we deduce that $b_t \geq a_t + (t+1)na_{t+1}$. Since $a_{t+1} > 0$, we may choose some n sufficiently large so that $b_t \geq 3$.

Finally, we obtain the desired result by iterating the above process, with $P(x+n)$ in place of $P(x)$, until $b_i \geq 3$ for all $i < q$ (a finite number of iterations). Note that if $Q(x) = P(x+n_1)$, then $Q(x+n_2) = P(x+n_3)$, where $n_3 = n_1 + n_2$. ∎

We are now ready to prove the claim. It is obviously true in the case where P is constant. If $P(x) = mx+p$ with $m \notin \{-1, 0, 1\}$, we just choose the arithmetic progression $u_n = mn + p - 1$. From now on, we suppose that P has degree at least 2.

Let $P(x) = \sum_{i=0}^{q} a_i x^i$, with $q \geq 2$ and $a_q > 0$. (The case $a_q < 0$ may be obtained by considering the polynomial $-P(x)$.) For any integer n, the sets $\{P(k) \mid k \in \mathbb{Z}\}$ and $\{P(k+n) \mid k \in \mathbb{Z}\}$ are equal. Therefore, by the lemma, it suffices to prove the result when $a_i \geq 3$ for all $i < q$.

Let $S = \sum_{\substack{i=0 \\ i \text{ odd}}}^{q} a_i$. Thus, $S \geq a_1 \geq 3$. Let k be any integer, and let $k = pS + r$ with $r \in \{0, 1, \ldots, S-1\}$. Then

$$P(k) = \sum_{i=0}^{q} a_i(pS+r)^i \equiv \sum_{i=0}^{q} a_i r^i \pmod{S} = P(r).$$

Therefore, the number $P(k)$ is congruent, modulo S, to one of the numbers $P(0), P(1), \ldots, P(S-1)$. Moreover,

$$P(S-1) = \sum_{i=0}^{q} a_i(S-1)^i \equiv \sum_{i=0}^{q} a_i(-1)^i \pmod{S} = P(1) - 2S$$
$$\equiv P(1) \pmod{S}.$$

Therefore, the set $\{P(k) \bmod S \mid k \in \mathbb{Z}\}$ contains at most $S-1$ elements.

Consequently, there exists $a \in \{0, 1, \ldots, S-1\}$ such that, for each integer k, we have $P(k) \not\equiv a \pmod{S}$. Then none of the terms of the arithmetic progression $u_n = nS + a$ belongs to the set $\{P(k) \mid k \in \mathbb{Z}\}$, and we are done.

St. Petersburg Mathematical Contest

The sum of two continuous periodic functions is a non-constant continuous periodic function. Prove that the periods of these two functions are integral multiples of the period of their sum.

Solution

The result is false.

Let $f(x) = \cos x - \cos 2x$ and $g(x) = \cos 2x$. Then f and g are continuous periodic functions, with minimal periods $T_f = 2\pi$ and $T_g = \pi$, respectively. Let $h = f + g$. Then $h(x) = \cos x$, which is a non-constant continuous periodic function, with minimal period $T_h = 2\pi$.

Since $T_g \neq kT_h$ for some integer k, the conclusion follows.

St. Petersburg Mathematical Contest

45

Let $P(z)$ and $Q(z)$ be complex polynomials, one of which is not constant. Every root of $P(z)$ is also a root of $Q(z)$ and vice versa. Every root of $P(z) - 1$ is also a root of $Q(z) - 1$ and vice versa. Prove that $P = Q$.

Solution

Suppose that $P(z)$ has degree $m \geq 1$. Since any root of $P(z)$ is also a root of $Q(z)$, we have $\deg Q(z) \geq 1$ as well. Without loss of generality, we will assume that $m \geq \deg Q(z)$. Let $u_1, u_2, \ldots, u_k$ be the distinct complex roots of $P(z)$, with respective multiplicities $r_1, r_2, \ldots, r_k$ (so that $r_1 + r_2 + \cdots + r_k = m$). Likewise, let $v_1, v_2, \ldots, v_\ell$ be the distinct complex roots of $P(z) - 1$, with respective multiplicities $s_1, s_2, \ldots, s_\ell$.

For $j = 1, 2, \ldots, k$ and $i = 1, 2, \ldots, \ell$, we have $P(u_j) = 0$ and $P(v_i) = 1$. Therefore, each of the u_j's is different from each of the v_i's. It follows that the derivative $P'(z)$, which is divisible by each $(z - u_j)^{r_j - 1}$ and by each $(z - v_i)^{s_i - 1}$, is divisible by the product

$$\prod_{j=1}^{k}(z - u_j)^{r_j - 1} \prod_{i=1}^{\ell}(z - v_i)^{s_i - 1}.$$

The latter has degree

$$(r_1 - 1) + \cdots + (r_k - 1) + (s_1 - 1) + \cdots + (s_\ell - 1) = 2m - (k + \ell),$$

and $P'(z)$ has degree $m - 1$. Thus, $m - 1 \geq 2m - (k + \ell)$; that is, $k + \ell \geq m + 1$.

This said, consider $R(z) = P(z) - Q(z) = (P(z) - 1) - (Q(z) - 1)$. The degree of $R(z)$ is at most m and $u_1, \ldots, u_k, v_1, \ldots, v_\ell$ are $k + \ell$ distinct roots of $R(z)$. Since $k + \ell \geq m + 1$, we must have $R = 0$; that is, $P = Q$.

Ukranian Mathematical Olympiad, 1999 (10th Grade)

Let $P(x)$ be a polynomial with integer coefficients. The sequence of integers x_1, x_2, ..., x_n, ... satisfies the conditions $x_1 = x_{2000} = 1999$, $x_{n+1} = P(x_n)$, $n \geq 1$. Find the value of

$$\frac{x_1}{x_2} + \frac{x_2}{x_3} + \cdots + \frac{x_{1999}}{x_{2000}}.$$

Solution

Subscripts are considered modulo 1999. For any positive integer n, let $y_n = x_n - x_{n-1}$. Then

$$\sum_{i=1}^{1999} y_i = \sum_{i=1}^{1999}(x_i - x_{i-1}) = x_{1999} - x_0 = 0. \quad (1)$$

Suppose that for all n, we have $y_n \neq 0$. Since $P(x)$ has integer coefficients, it is well known that, for any integers $a \neq b$, the integer $a - b$ divides $P(a) - P(b)$. It follows that y_n divides y_{n+1}, for all n. Then the numbers $|y_1|, |y_2|, \ldots, |y_n|, \ldots$ form a non-decreasing sequence. Since $|y_1| = |x_1 - x_{1999}| = |x_{2000} - x_{1999}| = |y_{2000}|$, we deduce that $|y_1| = |y_2| = \cdots = |y_{2000}|$. Let $a \neq 0$ be this common value.

Let k be the number of terms among $y_1, y_2, \ldots, y_{1999}$ which have the value a. Then the remaining $1999 - k$ terms have the value $-a$. Hence,

$$\sum_{i=1}^{1999} y_i = a(2k - 1999) \neq 0,$$

contradicting (1).

It follows that there is some n for which $y_n = 0$; that is, $x_n = x_{n-1}$. An easy induction leads to $x_n = x_1 = 1999$ for all n. Then

$$\frac{x_1}{x_2} + \frac{x_2}{x_3} + \cdots + \frac{x_{1999}}{x_{2000}} = 1 + 1 + \cdots + 1 = 1999.$$

XLIII Mathematical Olympiad of Moldova, 1999 (10th Form)

Let the function $f : \mathbb{R} \to \mathbb{R}$, $f(x) = x^2 - 2ax - a^2 - \frac{3}{4}$, be considered. Find the values a for which the inequality $|f(x)| \leq 1$ is true for every $x \in [0, 1]$.

Solution

We will prove that the values of a for which the inequality $|f(x)| \leq 1$ is true for every $x \in [0, 1]$ are those such that

$$-\tfrac{1}{2} \leq a \leq \tfrac{1}{2\sqrt{2}}. \qquad (1)$$

We will need the following function values:

$$f(0) = -a^2 - \tfrac{3}{4}, \qquad f(1) = -a^2 - 2a + \tfrac{1}{4}, \qquad f(a) = -2a^2 - \tfrac{3}{4}.$$

If $|a| > \tfrac{1}{2}$, then $|f(0)| = a^2 + \tfrac{3}{4} > \left(\tfrac{1}{2}\right)^2 + \tfrac{3}{4} = 1$, and therefore it is not true that $|f(x)| \leq 1$ for all $x \in [0, 1]$.

If $\tfrac{1}{2\sqrt{2}} < a \leq 1$, then $a \in [0, 1]$, and

$$|f(a)| = 2a^2 + \tfrac{3}{4} > 2\left(\tfrac{1}{2\sqrt{2}}\right)^2 + \tfrac{3}{4} = 1.$$

Thus, once again, it is not true that $|f(x)| \leq 1$ for all $x \in [0, 1]$.

The only values of a remaining to be considered are those satisfying (1). We will prove that, for all such a, we have $|f(x)| \leq 1$ for all $x \in [0, 1]$.

Note that $f(x) = (x - a)^2 + f(a)$. The graph of f is a parabola with vertex at $(a, f(a))$, opening upward. It follows that the maximum and minimum values of $f(x)$ on the interval can occur only where $x = 0$ or $x = 1$, or where $x = a$ if $a \in [0, 1]$. Therefore, the maximum value of $|f(x)|$ on $[0, 1]$ can occur only at these points.

Consider any a satisfying (1). Then, since $|a| \leq \tfrac{1}{2}$, we have

$$|f(0)| = a^2 + \tfrac{3}{4} \leq \left(\tfrac{1}{2}\right)^2 + \tfrac{3}{4} = 1.$$

Since $a \geq -\tfrac{1}{2}$, we have

$$1 - f(1) = a^2 + 2a + \tfrac{3}{4} = \left(a + \tfrac{3}{2}\right)\left(a + \tfrac{1}{2}\right) \geq 0,$$

and therefore $f(1) \leq 1$. Also, since $-\tfrac{5}{2} < a < \tfrac{1}{2}$, we get

$$f(1) - (-1) = -a^2 - 2a + \tfrac{5}{4} = \left(\tfrac{5}{2} + a\right)\left(\tfrac{1}{2} - a\right) > 0,$$

and therefore, $f(1) > -1$. Thus, $|f(1)| \leq 1$.

Finally, if $a \in [0, 1]$ and a satisfies (1), then

$$|f(a)| = 2a^2 + \tfrac{3}{4} \leq 2\left(\tfrac{1}{2\sqrt{2}}\right)^2 + \tfrac{3}{4} = 1.$$

We conclude that $|f(x)| \leq 1$ for all $x \in [0, 1]$.

XLIII Mathematical Olympiad of Moldova, 1999 (10th Form)

Find all the functions $f : \mathbb{R} \to \mathbb{R}$, which satisfy the relation
$$x \cdot f(x) = \lfloor x \rfloor \cdot f(\{x\}) + \{x\} \cdot f(\lfloor x \rfloor), \quad \forall\, x \in \mathbb{R},$$
where $\lfloor \cdot \rfloor$ and $\{\cdot\}$ denote the integral part and fractional part functions, respectively.

Solution

The solutions are the constant functions.

Any constant function satisfies the given relation, since $x = \lfloor x \rfloor + \{x\}$. Conversely, consider a function f for which the the given relation holds, and let $C = f(0)$. Substituting an arbitrary non-zero integer k for x in the given relation, we get $kf(k) = k \cdot C + 0 \cdot f(k) = k \cdot C$. Hence, $f(k) = C$. Similarly, for all $x \in (0,1)$, we have $xf(x) = 0 \cdot f(x) + x \cdot f(0) = C \cdot x$. Thus, $f(x) = C$.

Now let x be an arbitrary non-zero real number. Observing that $\lfloor x \rfloor \in \mathbb{Z}$ and $\{x\} \in [0,1)$, we get
$$xf(x) = \lfloor x \rfloor \cdot C + \{x\} \cdot C = C \cdot (\lfloor x \rfloor + \{x\}) = C \cdot x,$$
and $f(x) = C$ follows. In conclusion, $f(x) = C$ for all real numbers x.

XLIII Mathematical Olympiad of Moldova, 1999 (10th Form)

Find a polynomial of degree 3 with real coefficients such that each of its roots is equal to the square of one root of the polynomial $P(X) = X^3 + 9X^2 + 9X + 9$.

Solution

Let u, v, w be the (complex) roots of $P(X)$. We can obtain the required polynomial by two methods:

Solution I. Since $u + v + w = -9$, $uv + vw + wu = 9$, and $uvw = -9$,
$$u^2 + v^2 + w^2 = (u + v + w)^2 - 2(uv + vw + wu) = 63,$$
$$u^2v^2 + v^2w^2 + w^2u^2 = (uv + vw + wu)^2 - 2uvw(u + v + w) = -81,$$
$$u^2v^2w^2 = (uvw)^2 = 81.$$

Thus, the required polynomial is
$$Q(X) = (X - u^2)(X - v^2)(X - w^2) = X^3 - 63X^2 - 81X - 81.$$

Solution II. The polynomial $Q(X) = (X - u^2)(X - v^2)(X - w^2)$ satisfies
$$\begin{aligned}Q(X^2) &= (X^2 - u^2)(X^2 - v^2)(X^2 - w^2)\\ &= (X - u)(X - v)(X - w)(X + u)(X + v)(X + w)\\ &= P(X) \cdot (-P(-X))\\ &= (X^3 + 9X)^2 - (9X^2 + 9)^2 = X^6 - 63X^4 - 81X^2 - 81.\end{aligned}$$

Hence, $Q(X) = X^3 - 63X^2 - 81X - 81$.

XLIII Mathematical Olympiad of Moldova, 1999 (11th Form)

Let the number $n \in \mathbb{N}^*$ be given. Denote by M the set of all real numbers x for which there exists a finite sequence (a_p), $p = 1, \ldots, n$, with $a_p \in \{0, 1\}$, $p = 1, \ldots, n$, such that

$$x = 2^{-1} \cdot a_1 + 2^{-2} \cdot a_2 + \cdots + 2^{-n} \cdot a_n.$$

(a) Determine the set M, and prove that for every number $x \in M$ there exists a unique finite sequence (a_p), $p = 1, \ldots, n$, with the mentioned property.

(b) Find the function $f : M \to \mathbb{R}$ such that if (a_p) is the sequence defined above by the number x, then

$$f(x) = 2^{-1} \cdot 2000^{a_1} + 2^{-2} \cdot 2000^{a_2} + \cdots + 2^{-n} \cdot 2000^{a_n}, \quad \forall x \in M.$$

Solution

(a) Let $M' = \{2^n x : x \in M\}$. Then M' consists of all real numbers y that can be represented in the form

$$y = a_n + 2a_{n-1} + \cdots + 2^{n-1} a_1,$$

where $a_p \in \{0, 1\}$ for $p = 1, 2, \ldots, n$. The above representation for y is simply the binary expansion of y. Thus, $M' = \{0, 1, \ldots, 2^n - 1\}$ and therefore, $M = \{2^{-n} y : y \in M'\}$. The uniqueness of the representation for each $x \in M$ follows from the uniqueness of the binary expansion of each $y \in M'$.

(b) Let $x = 2^{-1} a_1 + 2^{-2} a_2 + \cdots + 2^{-n} a_n \in M$. Let $I = \{p \mid a_p \neq 0\}$. Then $x = \sum_{p \in I} 2^{-p}$. It follows that

$$\sum_{p \notin I} 2^{-p} = \sum_{p=1}^{n} 2^{-p} - \sum_{p \in I} 2^{-p} = 1 - 2^{-n} - x.$$

We deduce that

$$f(x) = \sum_{p \in I} \frac{2000}{2^p} + \sum_{p \notin I} \frac{1}{2^p} = 2000x + 1 - 2^{-n} - x$$
$$= 1999x + 1 - 2^{-n}.$$

Italian Team Selection Test, 1999

(a) Determine all the strictly monotone functions $f : \mathbb{R} \to \mathbb{R}$ such that
$$f(x + (f(y))) = f(x) + y, \quad \forall\, x, y \in \mathbb{R}.$$

(b) Prove that for every integer $n > 1$ there do not exist strictly monotone functions $f : \mathbb{R} \to \mathbb{R}$ such that
$$f(x + f(y)) = f(x) + y^n, \quad \forall\, x, y \in \mathbb{R}.$$

Solution

(a) Clearly, the functions $x \mapsto x$ and $x \mapsto -x$ meet the required conditions. We will show that there are no other solutions. We will use the notation f^2 for $f \circ f$, the composition of f with itself.

Let $f : \mathbb{R} \to \mathbb{R}$ be strictly monotone and have the property that, for all $x, y \in \mathbb{R}$,
$$f(x + f(y)) = f(x) + y. \tag{1}$$
With $x = y = 0$, equation (1) yields $f(f(0)) = f(0)$. Then, $f(0) = 0$, since f is injective.

First, suppose that f is strictly increasing. Note that the following property holds: if $f^2(a) = a$, then $f(a) = a$ (since $f(a) < a$ leads to the contradiction $a = f^2(a) < f(a)$ and, similarly, $f(a) > a$ is impossible).

Now, for arbitrary real numbers x, y, let $a = x + f(y)$. Using (1) twice,
$$f^2(a) = f(y + f(x)) = f(y) + x = a,$$
from which $f(a) = a$ follows. Using (1) again, this gives $f(x) + y = x + f(y)$, or $f(x) - x = f(y) - y$. Thus, the function $x \mapsto f(x) - x$ is constant. The constant must be 0, since $f(0) = 0$, and finally $f(x) = x$ for all x.

Next, suppose that f is strictly decreasing. For all x, y, the relation (1) provides
$$f(x + f^2(y)) = f(x) + f(y). \tag{2}$$
Hence, $f^2(x + f^2(y)) = f(f(x) + f(y)) = f^2(x) + y$. We deduce that the function f^2 satisfies condition (1), for all $x, y \in \mathbb{R}$. In addition, this function is strictly increasing. Thus, by the previous case, $f^2(x) = x$ for all x.

Returning to (2), we get the usual Cauchy functional equation
$$f(x + y) = f(x) + f(y), \quad \text{for all } x, y.$$
A well-known method gives first $f(k) = kf(1)$ for all integers k, next $f(r) = rf(1)$ for all rational r, and finally, $f(x) = xf(1)$ for all real x (using two adjacent rational sequences converging to x). In particular, for $x = f(1)$, we have $f^2(1) = (f(1))^2$. Substituting $x = 0$ and $y = 1$ into (1), we obtain $f^2(1) = f(0) + 1 = 1$. Then $(f(1))^2 = 1$. Observe that $f(1)$ is negative, since f is strictly decreasing and $f(0) = 0$. Therefore, $f(1) = -1$, and $f(x) = -x$ for all x.

(b) Suppose that f is strictly monotone and satisfies
$$f(x+f(y)) = f(x)+y^n,$$
for all $x,y \in \mathbb{R}$, where $n > 1$ is an integer. Since $f(0) = 0$ still holds, we have $f^2(y) = y^n$ for all y. Hence, for all y,
$$f(f^2(y)) = f^2(f(y)) = (f(y))^n.$$
For arbitrary x, y, we have, on the one hand,
$$f^2(x+f(y)) = (x+f(y))^n$$
and, on the other hand,
$$\begin{aligned} f^2(x+f(y)) &= f(f(x)+y^n) = f(y^n)+x^n \\ &= f(f^2(y))+x^n = (f(y))^n + x^n. \end{aligned}$$
Thus, for all x and y,
$$(x+f(y))^n = x^x + (f(y))^n. \qquad (3)$$
Since f is strictly increasing and $f(0) = 0$, there is some y such that $f(y) > 0$. Choose some such y and choose $x = 1$ in (3). Letting $b = f(y)$, we have $(1+b)^n = 1+b^n$, with $b > 0$. This is impossible for $n > 1$. The required result follows.

Estonian Mathematical Contest, 1996

Prove that the polynomial $P_n(x) = 1 + x + \frac{x^2}{2} + \frac{x^3}{6} + \cdots + \frac{x^n}{n!}$ has no zeroes if n is even and has exactly one zero if n is odd.

Solution

Note that $P_n(x) \geq 1$ for all $x \geq 0$. Therefore, if P_n has a zero ρ, then $\rho < 0$.

Let $F_n(x) = P_n(x)e^{-x}$. Then F_n has the same zeroes as P_n, and

$$F_n'(x) = (P_n'(x) - P_n(x))e^{-x} = -\frac{x^n}{n!}e^{-x}. \qquad (1)$$

If n is even, then we see from (1) that F_n is decreasing on $\mathbb{R}$. Since $F_n(0) = P_n(0) = 1$, we have $F_n(x) \geq 1$ for all $x \leq 0$. Therefore, F_n has no negative zeroes, and hence, neither does P_n. Thus, P_n has no zeroes at all.

Now assume that n is odd. Then $P_n(x)$, being a polynomial of odd degree, has at least one real zero. From (1), the function F_n is strictly decreasing on the interval $(-\infty, 0]$. Hence, F_n has at most one zero in this interval. It follows that P_n has at most one zero.

Finally, we prove that if P_n has a real zero ρ, then the multiplicity of ρ must be 1. Suppose instead that P_n has a zero ρ of multiplicity $k > 1$. Then $P_n(x) = (x-\rho)^k Q_n(x)$, for some polynomial $Q_n(x)$, and hence,

$$P_n'(x) = k(x-\rho)^{k-1} Q_n(x) + (x-\rho)^k Q_n'(x).$$

Now $P_n(\rho) = P_n'(\rho) = 0$. Then $0 = P_n(\rho) - P_n'(\rho) = \frac{\rho^n}{n!}$, which implies that $\rho = 0$. But $\rho < 0$, since ρ is a zero of P_n. We have a contradiction.

Mongolian Team Selection Test for 40th IMO, 1999

Let n be a positive integer and $P(x)$ a polynomial of degree $2n$ such that $P(0) = 1$ and $P(k) = 2^{k-1}$ for $k = 1, 2, \ldots, 2n$. Prove that $2P(2n+1) - P(2n+2) = 1$.

Solution

For $k = 0, 1, \ldots, 2n$, let
$$L_k(x) = \prod_{\substack{j=0 \\ j \neq k}}^{2n} (x-j).$$

The polynomial $\sum_{k=0}^{2n} \dfrac{P(k)}{L_k(k)} L_k(x)$ has degree at most $2n$ and takes the same values as $P(x)$ at $0, 1, \ldots, 2n$. Thus,
$$P(x) = \sum_{k=0}^{2n} \frac{P(k)}{L_k(k)} L_k(x). \qquad (1)$$

Let $P(x) = a_{2n} x^{2n} + a_{2n-1} x^{2n-1} + \cdots + a_0$. We use (1) to calculate the coefficient a_{2n}:

$$\begin{aligned}
a_{2n} &= \sum_{k=0}^{2n} \frac{P(k)}{L_k(k)} = \sum_{k=0}^{2n} \frac{P(k)}{(-1)^k k!(2n-k)!} \\
&= \frac{1}{(2n)!} \sum_{k=0}^{2n} (-1)^k \binom{2n}{k} P(k) \\
&= \frac{1}{(2n)!} \left(\binom{2n}{0} 1 - \binom{2n}{1} 1 + \binom{2n}{2} 2^1 + \cdots + \binom{2n}{2n} 2^{2n-1} \right) \\
&= \frac{1}{2} \frac{1}{(2n)!} \left(1 + \sum_{k=0}^{2n} (-1)^k \binom{2n}{k} 2^k \right) \\
&= \frac{1}{2} \frac{1}{(2n)!} (1 + (1-2)^{2n}) = \frac{1}{(2n)!}.
\end{aligned}$$

Now, consider $Q(x) = 2P(x) - P(x+1)$. The polynomial $Q(x)$ has the same degree and the same leading coefficient as $P(x)$. From the hypotheses, we must have $Q(0) = 1$ and $Q(k) = 0$ for $k = 1, 2, \ldots, 2n-1$. It follows that
$$Q(x) = a_{2n}(x-1)(x-2)\cdots(x-(2n-1))(x-r)$$
for some real number r. Since $Q(0) = 1$ and $a_{2n} = \dfrac{1}{(2n)!}$, we find that
$$1 = \frac{(-1)^{2n} r}{(2n)!} (2n-1)! = \frac{r}{2n}; \text{ hence, } r = 2n.$$

Thus, $Q(x) = \dfrac{1}{(2n)!} (x-1) \cdots (x-2n)$. Then
$$2P(2n+1) - P(2n+2) = Q(2n+1) = \frac{(2n)!}{(2n)!} = 1.$$

Korean Mathematical Olympiad, 2000

Determine all functions f from the set of real numbers to itself such that for every x and y,
$$f(x^2 - y^2) = (x - y)(f(x) + f(y)).$$

Solution

The solution is the set of linear functions passing through the origin.

Let f be a function satisfying the given functional equation. Setting $x = y = 0$ in this equation leads to $f(0) = 0$. Then, setting $y = -x$ gives $0 = f(0) = 2x(f(x) + f(-x))$, from which we deduce that f is an odd function.

For all x and y, we have
$$f(x^2 - y^2) = (x - y)(f(x) + f(y))$$
and also (replacing y by $-y$ and noting that $f(-y) = -f(y)$),
$$f(x^2 - y^2) = (x + y)(f(x) - f(y)).$$
Hence,
$$(x - y)(f(x) + f(y)) = (x + y)(f(x) - f(y));$$
that is, $xf(y) = yf(x)$. Setting $y = 1$, we get $f(x) = xf(1)$. Thus, f is linear.

Conversely, it is easy to verify that any linear function passing through the origin is a solution of the problem.

Vietnamese Mathematical Olympiad, 2000

For every integer $n \geq 3$ and any given angle α in $(0, \pi)$, let $P_n(x) = x^n \sin \alpha - x \sin n\alpha + \sin(n-1)\alpha$.

(a) Prove that there is only one polynomial of the form $f(x) = x^2 + ax + b$ such that for every $n \geq 3$, $P_n(x)$ is divisible by $f(x)$.

(b) Prove that there does not exist a polynomial $g(x)$ of the form $g(x) = x + c$ such that for every $n \geq 3$, $P_n(x)$ is divisible by $g(x)$.

Solution

First we observe that

$$P_{n+1}(x) - xP_n(x)$$
$$= x^2 \sin n\alpha - x(\sin(n+1)\alpha + \sin(n-1)\alpha) + \sin n\alpha$$
$$= x^2 \sin n\alpha - 2x \sin n\alpha \cos \alpha + \sin n\alpha$$
$$= (\sin n\alpha)(x^2 - 2x \cos \alpha + 1).$$

(a) Let $f(x) = x^2 + ax + b$, and suppose that $f(x)$ divides $P_n(x)$ for all $n \geq 3$. Then $f(x)$ divides $P_{n+1}(x) - P_n(x) = (\sin n\alpha)(x^2 - 2x \cos \alpha + 1)$ for all $n \geq 3$. Choosing n such that $\sin n\alpha \neq 0$ ($n = 3$ or $n = 4$ will do), we see that necessarily $f(x) = x^2 - 2x \cos \alpha + 1$.

Conversely, since

$$\begin{aligned} P_n(x) &= xP_{n-1}(x) + \big(\sin(n-1)\alpha\big)f(x) \\ P_{n-1}(x) &= xP_{n-2}(x) + \big(\sin(n-2)\alpha\big)f(x) \\ &\vdots \\ P_2(x) &= 0 + (\sin \alpha)f(x), \end{aligned}$$

we have $P_n(x) = f(x)\big(\sin(n-1)\alpha + x\sin(n-2)\alpha + \cdots + x^{n-2} \sin \alpha\big)$. Thus, $f(x)$ divides every $P_n(x)$ with $n \geq 3$.

(b) Suppose $g(x) = x + c$ divides $P_n(x)$ for all $n \geq 3$. In particular, $g(x)$ divides $P_3(x)$. From (a), we have $P_3(x) = (\sin \alpha)f(x)(x + 2 \cos \alpha)$. If we assume that c is a real number, then $g(x)$ does not divide $f(x)$, since $f(x)$ has two non-real roots, $e^{i\alpha}$ and $e^{-i\alpha}$. Therefore, $g(x)$ must divide $x + 2\cos \alpha$, and we must have $c = -2\cos \alpha$.

Since $g(x)$ divides $P_4(x)$, it follows that $P_4(-2\cos \alpha) = 0$, which yields $3 - 4\sin^2 \alpha = 0$. Then $\alpha = \frac{\pi}{3}$ or $\alpha = \frac{2\pi}{3}$. But in each of these cases, $-2\cos \alpha$ is not a root of $P_5(x)$. Thus, $g(x) = x + c$ cannot divide all $P_n(x)$ (if c is real).

Bulgarian Mathematical Olympiad, 2000

Find all polynomials $P(x)$ with real coefficients such that we have $P(x)P(x+1) = P(x^2)$ for all real x.

Solution

If $P \equiv 0$ or $P \equiv 1$, then P is clearly a solution. These are the only constant polynomials which are solutions of the problem.

Now suppose that P is a non-constant polynomial which is a solution. Then $P(x)P(x+1) = P(x^2)$ for all real x. Hence, $P(z)P(z+1) = P(z^2)$ for all $z \in \mathbb{C}$. Let $z \in \mathbb{C}$ be a root of P. Then, $P(z^2) = 0$ and $P((z-1)^2) = 0$.

Suppose that $0 < |z| < 1$. Define a sequence $\{z_n\}_{n=0}^{\infty}$ by $z_0 = z$ and $z_{n+1} = z_n^2$ for $n \geq 0$. Then, for each $n \geq 0$, we have $0 < |z_{n+1}| < |z_n| < 1$ and $P(z_n) = 0$. Thus, P has an infinite number of distinct roots, and then $P \equiv 0$, a contradiction.

Similarly, if $|z| > 1$, then $|z_{n+1}| > |z_n| > 1$ and $P(z_n) = 0$, which leads to the same contradiction.

Now suppose that $|z| = 1$ and $z \notin \{1, e^{i\frac{\pi}{3}}, e^{-i\frac{\pi}{3}}\}$. Let $z = e^{i\theta}$. Then $|(z-1)^2| = 2(1-\cos\theta) \in (0,1) \cup (1,4)$. Considering the sequence defined by $z_0 = (1-z)^2$ and $z_{n+1} = z_n^2$ for $n \geq 0$, we use an argument like the one above to get a contradiction.

Thus, the only possible roots of P are 0, 1, $e^{i\frac{\pi}{3}}$, and $e^{-i\frac{\pi}{3}}$. Since P has real coefficients, we have $P(e^{i\frac{\pi}{3}}) = 0$ if and only if $P(e^{-i\frac{\pi}{3}}) = 0$. If $e^{i\frac{\pi}{3}}$ is a root, then $P(e^{i\frac{\pi}{6}})P(e^{i\frac{\pi}{6}}+1) = P(e^{i\frac{\pi}{3}}) = 0$, and $e^{i\frac{\pi}{6}}$ or $e^{i\frac{\pi}{6}}+1$ is a root of P. Since these numbers do not belong to $\{0, 1, e^{i\frac{\pi}{3}}, e^{-i\frac{\pi}{3}}\}$, we have a contradiction.

Now the only possible roots are 0 and 1. Thus, $P(x) = ax^p(x-1)^q$, where a is a non-zero real number, and p and q are non-negative integers. Then $P(x)P(x+1) = P(x^2)$ is equivalent to

$$a^2 x^{p+q}(x-1)^p(x+1)^q = ax^{2p}(x-1)^q(x+1)^q.$$

It follows that $a = 1$ and $p = q$. Thus, $P(x) = x^p(x-1)^p$. The integer p must now be positive, since P is supposed to be non-constant.

In conclusion, the solutions are $P(x) = 0$ and $P(x) = x^p(x-1)^p$, where p may be any non-negative integer.

Bulgarian Mathematical Olympiad, 2000

6

Let $\mathcal{A}$ be the set of all binary sequences of length n, and let $0 \in \mathcal{A}$ be the sequence all terms of which are zeroes. The sequence $c = \langle c_1, c_2, \ldots, c_n \rangle$ is called the sum of $a = \langle a_1, a_2, \ldots, a_n \rangle$ and $b = \langle b_1, b_2, \ldots, b_n \rangle$ if $c_i = 0$ when $a_i = b_i$ and $c_i = 1$ when $a_i \neq b_i$. Let $f : \mathcal{A} \to \mathcal{A}$ be a function such that $f(0) = 0$ and if the sequences a and b differ in exactly k terms then the sequences $f(a)$ and $f(b)$ differ also exactly in k terms. Prove that if a, b, and c are sequences from $\mathcal{A}$ such that $a + b + c = 0$, then $f(a) + f(b) + f(c) = 0$.

Solution

The set $\mathcal{A}$ is a vector space on $\mathbb{Z}/2\mathbb{Z}$ (the sum is the one defined in the statement, and let $0 \cdot a = 0$ and $1 \cdot a = a$) with dimension n and canonical basis $B_1 = (e_1, \ldots, e_n)$, where $e_1 = (1, 0, \ldots, 0)$, $e_2 = (0, 1, 0, \ldots, 0)$, $\ldots$, $e_n = (0, \ldots, 0, 1)$. If $a = \langle a_1, a_2, \ldots, a_n \rangle$, then $a = \sum_{i=1}^{n} a_i e_i$.

Let $d(a, b) = \sum_{i=1}^{n} |a_i - b_i|$. Then $d(a, b)$ is simply the number of terms that differ in the sequences a and b. From the statement of the problem, $d(f(a), f(b)) = d(a, b)$. It follows trivially that f is injective.

Let $i \in \{1, \ldots, n\}$. Since $f(0) = 0$ and $d(0, e_i) = 1$, we deduce that $d(0, f(e_i)) = 1$. It follows that there exists $m \in \{1, \ldots, n\}$ such that $f(e_i) = e_m$. Let $f_i = f(e_i)$. Since f is injective, we deduce that $B_2 = (f_1, \ldots, f_n)$ is a permutation of $(e_1, \ldots, e_n)$ and, therefore, is also a basis for $\mathcal{A}$.

Let $a = (a_1, a_2, \ldots, a_n)_{B_1}$ with $d(0, a) = k$. Then $d(0, f(a)) = k$. Let $i \in \{1, \ldots, n\}$.

- If $a_i = 1$, then $d(e_i, a) = k - 1$. Thus, $d(f_i, f(a)) = k - 1$; that is, the ith coordinate of $f(a)$ in B_2 is equal to 1.

- If $a_i = 0$, then $d(e_i, a) = k$. Thus, $d(f_i, f(a)) = k$; that is, the ith coordinate of $f(a)$ in B_2 is equal to 0.

It follows that, if $a = \sum_{i=1}^{n} a_i e_i$, then $f(a) = \sum_{i=1}^{n} a_i f_i = \sum_{i=1}^{n} a_i f(e_i)$. Thus, f is linear.

Then, if $a + b + c = 0$, we have

$$f(a) + f(b) + f(c) = f(a + b + c) = f(0) = 0.$$

Taiwanese Mathematical Olympiad, 2000

Let f be a function from the set of positive integers to the set of non-negative integers such that $f(1) = 0$ and

$$f(n) = \max\{f(j) + f(n-j) + j\}$$

for all $n \geq 2$. Determine $f(2000)$.

Solution

In what follows, the maximum in the definition of $f(n)$ is considered over all j such that $1 \leq j \leq n-1$ (that is, all j for which $f(j)$ and $f(n-j)$ are defined). We will prove by induction that $f(n) = \dfrac{n(n-1)}{2}$ for all $n \geq 1$.

It is readily checked that $f(2) = f(1) + f(1) + 1 = 1$, that $f(3) = \max\{f(1) + f(2) + 1, f(2) + f(1) + 2\} = 3$, and that $f(4) = 6$. Assume that $n \geq 5$ and that $f(k) = \dfrac{k(k-1)}{2}$ for $1 \leq k < n$. Then

$$f(n-1) + f(1) + n - 1 = \frac{(n-2)(n-1)}{2} + 0 + n - 1 = \frac{n(n-1)}{2},$$

and for $1 \leq j \leq n-2$,

$$\begin{aligned} f(j) + f(n-j) + j &= \frac{j(j-1)}{2} + \frac{(n-j)(n-j-1)}{2} + j \\ &= \frac{n(n-1) - 2j(n-1-j)}{2} < \frac{n(n-1)}{2}. \end{aligned}$$

It follows that $f(n) = \max\{f(j) + f(n-j) + j\} = \dfrac{n(n-1)}{2}$. This concludes the induction.

Now, taking $n = 2000$ in the formula we have just proved, we find that $f(2000) = 1999000$.

Iranian Mathematical Olympiad, 2000

Prove that for every positive integer n, there exists a polynomial $p(x)$ with integer coefficients such that $p(1)$, $p(2)$, ..., $p(n)$ are distinct powers of 2.

Solution

We will prove a stronger statement: For every positive integer n, there exists a polynomial $p(x)$ with integer coefficients and degree at most n such that $p(0)$, $p(1)$, ..., $p(n)$ are distinct powers of 2.

Define an $(n+1) \times (n+1)$ matrix M as follows:

$$M = \begin{pmatrix} 1 & 0 & 0 & \cdots & 0 \\ 1 & 1 & 1 & \cdots & 1 \\ 1 & 2 & 2^2 & \cdots & 2^n \\ \vdots & \vdots & \vdots & \ddots & \vdots \\ 1 & n & n^2 & \cdots & n^n \end{pmatrix}.$$

For convenience, we will index the rows and columns of our matrices and vectors starting with 0 rather than 1. Then the entries of M are $m_{ij} = i^j$, for $i = 0, 1, 2, \ldots, n$ and $j = 0, 1, 2, \ldots, n$ (with $0^0 = 1$).

Let d denote the determinant of M. Since M is a Van der Monde matrix, we have $d = \prod_{0 \le i < j \le n} (i-j)$, by the well-known formula for a Van der Monde determinant. It follows that d is a non-zero integer and M is invertible. The entries of M^{-1} are of the form t_{ij}/d, where t_{ij} is an integer.

Since

$$M \begin{pmatrix} 1 \\ 0 \\ 0 \\ \vdots \\ 0 \end{pmatrix} = \begin{pmatrix} 1 \\ 1 \\ 1 \\ \vdots \\ 1 \end{pmatrix},$$

we must have

$$M^{-1} \begin{pmatrix} 1 \\ 1 \\ 1 \\ \vdots \\ 1 \end{pmatrix} = \begin{pmatrix} 1 \\ 0 \\ 0 \\ \vdots \\ 0 \end{pmatrix}.$$

Therefore,

$$\sum_{j=0}^{n} \frac{t_{ij}}{d} = \begin{cases} 1, & \text{if } i = 0, \\ 0, & \text{if } i > 0. \end{cases}$$

Note in particular that the above sum is an integer for all i.

Let $d = 2^a b$, where $a, b \ge 0$ are integers and b is odd. Let ω be the order of 2 in the ring of integers modulo b. Then $2^\omega \equiv 1 \pmod{b}$ and $\omega \ge 1$. For every integer $k \ge 0$, we have $2^{k\omega} \equiv 1 \pmod{b}$, and hence there exists an integer c_k such that $2^{k\omega} = 1 + c_k b$.

Let $a_0, a_1, \ldots, a_n$ be such that

$$\begin{pmatrix} a_0 \\ a_1 \\ a_2 \\ \vdots \\ a_n \end{pmatrix} = M^{-1} \begin{pmatrix} 2^a \\ 2^{a+\omega} \\ 2^{a+2\omega} \\ \vdots \\ 2^{a+n\omega} \end{pmatrix}.$$

Then, for all i,

$$a_i = \sum_{j=0}^{n} \frac{t_{ij}}{d} 2^{a+j\omega} = \sum_{j=0}^{n} \frac{t_{ij}}{d} 2^a (1 + c_j b) = 2^a \sum_{j=0}^{n} \frac{t_{ij}}{d} + \sum_{j=0}^{n} t_{ij} c_j.$$

Since $\sum_{j=0}^{n} \frac{t_{ij}}{d}$ is an integer, we see that a_i is an integer for each i.

Therefore, the polynomial $p(x) = a_n x^n + a_{n-1} x^{n-1} + \cdots + a_1 x + a_0$ has integer coefficients. Its degree is at most n, clearly. Moreover, since

$$M \begin{pmatrix} a_0 \\ a_1 \\ a_2 \\ \vdots \\ a_n \end{pmatrix} = \begin{pmatrix} 2^a \\ 2^{a+\omega} \\ 2^{a+2\omega} \\ \vdots \\ 2^{a+n\omega} \end{pmatrix},$$

we have $p(i) = 2^{a+i\omega}$ for each $i = 0, 1, \ldots, n$, and we are done.

Shortlist for IMO, 2000 (Belarus)

12. Find all pairs of functions f and g from the set of real numbers to itself such that $f(x+g(y)) = xf(y) - yf(x) + g(x)$ for all real numbers x and y.

Solution

It is easy to see that if f is constant then $f(x) = g(x) \equiv 0$. Assume that f is non-constant. Replacing x by $g(x)$ in the given equation, we have

$$f(g(x) + g(y)) = g(x)f(y) - yf(g(x)) + g(g(x)).$$

Since the left side is symmetric in x and y, the right side must be also, and we obtain

$$g(x)f(y) - yf(g(x)) + g(g(x)) = g(y)f(x) - xf(g(y)) + g(g(y)). \quad (1)$$

Taking $x = 0$ in the original equation we get

$$f(g(y)) = ay + b \quad (2)$$

where $a = -f(0)$ and $b = g(0)$. Setting $y = 0$ in (2) gives $f(b) = b$. Equation (1) with $y = 0$ now gives

$$g(g(x)) = ag(x) + bf(x) - bx + g(b). \quad (3)$$

Inserting (2) and (3) into (1), we obtain

$$g(x)f(y) + ag(x) + bf(x) = g(y)f(x) + ag(y) + bf(y),$$

or equivalently,

$$\bigl(g(x) - b\bigr)\bigl(f(y) + a\bigr) = \bigl(g(y) - b\bigr)\bigl(f(x) + a\bigr), \quad (4)$$

for all x and y.

Recalling that f is non-constant, we choose y_0 such that $f(y_0) + a \neq 0$. Then we see from (4) with $y = y_0$ that

$$g(x) - b = A(f(x) + a),$$

where $A = \dfrac{g(y_0) - b}{f(y_0) + a}$. Letting $B = Aa + b$, we have

$$g(x) = Af(x) + B, \quad (5)$$

for all $x \in \mathbb{R}$. Inserting this into the original equation, we obtain

$$f(x + g(y)) = xf(y) - yf(x) + Af(x) + B.$$

Putting $y = A$ gives $f(x + g(A)) = xf(A) + B$, which shows that f is a linear function. By (5), g is also linear. Now, a direct computation yields

$$f(x) = \frac{c}{c+1}(x - c)) \quad \text{and} \quad g(y) = c(y - c),$$

for any constant $c \neq -1$.

Shortlist for IMO, 2000 (United Kingdom)

A function F is defined from the set of non-negative integers to itself such that, for every non-negative integer n, $F(4n) = F(2n) + F(n)$, $F(4n + 2) = F(4n) + 1$, and $F(2n + 1) = F(2n) + 1$. Prove that, for each positive integer m, the number of integers n with $0 \leq n < 2^m$ and $F(4n) = F(3n)$ is $F(2^{m+1})$.

Solution

We have $F(0) = 0$ from the first given equation, and we see that F is uniquely determined by the given equations. The following table shows the values of $F(n)$ for $0 \leq n \leq 16$:

n	0	1	2	3	4	5	6	7	8	9	10	11	12	13	14	15	16
$F(n)$	0	1	1	2	2	3	3	4	3	4	4	5	5	6	6	7	5

From the first given equation, by induction, we obtain $F(2^k) = f_{k+1}$, for $k = 0, 1, 2, \ldots$, where $\{f_n\}_{n=1}^{\infty}$ is the Fibonacci sequence, defined by

$$f_1 = f_2 = 1, \qquad f_n = f_{n-1} + f_{n-2} \text{ for } n \geq 3.$$

In fact, a general formula for $F(n)$ may be given in terms of the Fibonacci numbers: if n has the binary representation

$$n = \varepsilon_k 2^k + \varepsilon_{k-1} 2^{k-1} + \cdots + \varepsilon_1 2 + \varepsilon_0,$$

where $\varepsilon_i \in \{0, 1\}$ for each i, then

$$F(n) = \varepsilon_k f_{k+1} + \varepsilon_{k-1} f_k + \cdots + \varepsilon_1 f_2 + \varepsilon_0 f_1. \qquad (1)$$

This formula may be verified simply by checking that it satisfies the given equations defining $F(n)$.

We now prove that $F(3n) \leq F(4n)$ for all non-negative integers n. We proceed by induction on m, the number of digits in the binary representation of n. When $m = 1$, we have $n = 0$ or $n = 1$, in both of which cases $F(3n) = F(4n)$, as can be seen from the table above.

Now let an integer $m \geq 2$ be fixed. As our induction hypothesis, we assume that $F(3n) \leq F(4n)$ whenever n has fewer than m digits in its binary representation. Consider any positive integer n which has exactly m digits in its binary representation. Thus, $n = 2^{m-1} + p$, where $0 \leq p < 2^{m-1}$. By the induction hypothesis, $F(3p) \leq F(4p)$. Using (1), we obtain

$$F(4n) = F(2^{m+1} + 4p) = f_{m+2} + F(4p) \geq f_{m+2} + F(3p). \qquad (2)$$

On the other hand

$$F(3n) = F(3 \cdot 2^{m-1} + 3p) = F(2^m + 2^{m-1} + 3p). \qquad (3)$$

To proceed further with the calculation of $F(3n)$, we consider three cases:

Case 1. $3p < 2^{m-1}$.

Then the number of digits in the binary representation of $3p$ is less than m, and the binary representation of $3p$ does not carry into that of $3 \cdot 2^{m-1}$.

Starting from (3), we can apply (1) and then (2) to get

$$F(3n) = f_{m+1} + f_m + F(3p) = f_{m+2} + F(3p) \leq F(4n). \quad (4)$$

Case 2. $2^{m-1} < 3p < 2^m$.

Then the number of digits in the binary representation of $3p$ is exactly m, and the binary representation of $3p$ carries 1 into that of $3 \cdot 2^{m-1}$. Starting from (3) and using (1), we calculate

$$\begin{aligned} F(3n) &= F(2^m + 2^{m-1} + 2^{m-1} + (3p - 2^{m-1})) \\ &= F(2^{m+1} + (3p - 2^{m-1})) \\ &= f_{m+2} + F(3p) - f_m = f_{m+1} + F(3p). \end{aligned}$$

Noting that $f_{m+1} < f_{m+2}$ (since $m \geq 2$) and using (2), we find that $F(3n) < F(4n)$.

Case 3. $2^m < 3p < 3 \cdot 2^{m-1} \; (= 2^m + 2^{m-1})$.

Then the number of digits in the binary representation of $3p$ is exactly $m + 1$, and the binary representation of $3p$ carries 10 into that of $3 \cdot 2^{m-1}$. Starting from (3) and using (1), we calculate

$$\begin{aligned} F(3n) &= F(2^m + 2^{m-1} + 2^m + (3p - 2^m)) \\ &= F(2^{m+1} + 2^{m-1} + (3p - 2^m)) \\ &= f_{m+2} + f_m + F(3p) - f_{m+1} = f_{m+2} - f_{m-1} + F(3p). \end{aligned}$$

Noting that $f_{m-1} > 0$ and using (2), we find that $F(3n) < F(4n)$.

In each case, we have shown that $F(3n) \leq F(4n)$. This completes the induction proof.

Let E denote the set consisting of 0 together with all positive integers in whose binary representation the 1s are isolated (meaning that no two 1s appear in adjacent positions). For example, $1 \in E$, $2 \in E$, but $3 \notin E$. We will now prove that $F(3n) = F(4n)$ if and only if $n \in E$.

First suppose that $n \in E$. If $n = 0$, then $F(3n) = F(0) = F(4n)$. Now assume that $n > 0$. On multiplying n by 3 (that is, 11 in binary), each 01 (and the leading 1) is replaced by 11. Consequently, $F(3n)$ is obtained from $F(n)$ by replacing each f_{i+1} in (1) by $f_{i+1} + f_{i+2} = f_{i+3}$. On the other hand, $F(4n)$ is obtained from $F(n)$ by replacing each f_{i+1} in (1) by f_{i+3}. Thus, $F(3n) = F(4n)$.

Conversely, suppose that n is a non-negative integer such that $F(3n) = F(4n)$. If $n = 0$ or $n = 1$, then $n \in E$. Now assume that $n \geq 2$. Let m be the number of digits in the binary representation of n. Upon examining the induction proof above, we observe that the equation $F(3n) = F(4n)$ occurs only in Case 1 and only when $F(3p) = F(4p)$. We note also that, in Case 1, we have $p < 2^{m-1}/3 < 2^{m-1}/2 = 2^{m-2}$, which implies that the second binary digit of n is 0. Therefore, in Case 1, we have $n \in E$ if and only if $p \in E$. By refining the induction proof slightly, we conclude that for all non-negative integers n, the equality $F(3n) = F(4n)$ implies that $n \in E$.

We have proved our claim that $F(3n) = F(4n)$ if and only if $n \in E$. Finally, we will prove that, for each non-negative integer m, the number of integers $n \in E$ such that $0 \leq n < 2^m$ is f_{m+2}. Since $f_{m+2} = F(2^{m+1})$, we will have proved the desired result.

We proceed by induction on m. For $m = 0$, the number of $n \in E$ such that $0 \leq n < 2^m$ is $1 = f_2$; for $m = 1$, the number of $n \in E$ such that $0 \leq n < 2^m$ is $2 = f_3$. Now fix an integer $m \geq 2$, and assume, as the induction hypothesis, that the number of $n \in E$ such that $0 \leq n < 2^k$ is f_{k+2} for all integers k with $0 \leq k < m$. Among the integers $n \in E$ such that $0 \leq n < 2^m$ are those such that $0 \leq n < 2^{m-1}$, the number of which is f_{m+1} (using the induction hypothesis). The remaining $n \in E$ such that $0 \leq n < 2^n$ are those such that $2^{m-1} \leq n < 2^m$. The first two binary digits of any such number n must be 10, and the remaining digits represent a number in E which is less than 2^{m-2}. Thus, the number of such n is f_n (using the induction hypothesis again). The total number of $n \in E$ such that $0 \leq n < 2^n$ is $f_{m+1} + f_m = f_{m+2}$. This completes the proof.

32nd Austrian Mathematical Olympiad

Determine all functions $f : \mathbb{R} \mapsto \mathbb{R}$, such that for all real numbers x and y the functional equation $f(f(x)^2 + f(y)) = x \cdot f(x) + y$ is satisfied.

Solution

Suppose f is a function $f : \mathbb{R} \to \mathbb{R}$ that satisfies
$$f(f(x)^2 + f(y)) = x \cdot f(x) + y. \tag{1}$$
By setting $x = 0$ in (1), we get
$$f(f(0)^2 + f(y)) = y.$$
Thus, f can take any value in $\mathbb{R}$; hence, it is surjective. Also, if $f(a) = f(b)$, then $a = f(f(0)^2 + f(a)) = f(f(0)^2 + f(b)) = b$. Therefore, f is injective. Thus, we see that f must be a bijective function.

Let h be a number such that $f(h) = 0$. By setting $x = y = h$ in (1), we get $f(0) = h$. Then, putting $x = y = 0$ in (1) gives $f(h^2 + h) = 0$. Since f is injective and $f(h^2 + h) = f(h)$, we must have $h^2 + h = h$; hence, $h = 0$. Thus, $f(0) = 0$.

By setting $x = 0$ in (1), we obtain
$$f(f(y)) = y. \tag{2}$$
And, by setting $y = 0$ in (1), we get
$$f(f(x)^2) = x \cdot f(x). \tag{3}$$
Replacing x by $f(x)$ in (3) gives $f\big(f(f(x))^2\big) = f(x)f(f(x))$. Using (2), we simplify this to $f(x^2) = x \cdot f(x)$. By comparing this with (3), we get $f(f(x)^2) = f(x^2)$. Hence (since f is injective), $f(x)^2 = x^2$ for all x.

Suppose that there exist a and b in $\mathbb{R} \setminus \{0\}$ such that $f(a) = a$ and $f(b) = -b$. Setting $x = a$ and $y = b$ in (1) gives $f(a^2 - b) = a^2 + b$, which does not obey $f(x)^2 = x^2$. Therefore, either $f(x) = x$ for all x, or $f(x) = -x$ for all x.

14th Nordic Mathematical Contest

4

The real-valued function f is defined for $0 \leq x \leq 1$, and satisfies $f(0) = 0$, $f(1) = 1$, and
$$\frac{1}{2} \leq \frac{f(z) - f(y)}{f(y) - f(x)} \leq 2,$$
for all $0 \leq x < y < z \leq 1$ with $z - y = y - x$. Prove that
$$\frac{1}{7} \leq f\left(\frac{1}{3}\right) \leq \frac{4}{7}.$$

Solution

Let $f\left(\frac{1}{3}\right) = a$ and $f\left(\frac{2}{3}\right) = b$. Setting $x = 0$, $y = \frac{1}{3}$, and $z = \frac{2}{3}$, we get
$$\frac{1}{2} \leq \frac{b-a}{a} \leq 2. \qquad (1)$$
Setting $x = \frac{1}{3}$, $y = \frac{2}{3}$, and $z = 1$, we obtain
$$\frac{1}{2} \leq \frac{1-b}{b-a} \leq 2. \qquad (2)$$

Suppose that $a < 0$. From (1), we deduce that $b - a < 0$; then $b < 0$, and hence $1 - b > 0$. Then, from (2), we get $b - a > 0$, a contradiction. Thus, $a > 0$.

Using (1), we deduce that $b - a > 0$. Then (1) can be rewritten as
$$b \leq 3a \leq 2b, \qquad (3)$$
and (2) can be rewritten as
$$1 + 2a \leq 3b \leq 2 + a, \qquad (4)$$
From the inequalities on the right in (3) and (4), we get $2 + a \geq 3b \geq \frac{9}{2}a$, and then $a \leq \frac{4}{7}$. From the inequalities on the left, we get $1 + 2a \leq 3b \leq 9a$, and then $a \geq \frac{1}{7}$. Thus, $\frac{1}{7} \leq a \leq \frac{4}{7}$, and we are done.

Ukrainian Mathematical Olympiad, 2001 (Grade 11)

Does there exist a function $f : \mathbb{R} \to \mathbb{R}$ such that for all x, $y \in \mathbb{R}$ the following equality holds?

$$f(xy) = \max\{f(x), y\} + \min\{f(y), x\}.$$

Solution

Suppose that there exists such a function. Then, setting $x = y = 1$ in the given equation, we get

$$f(1) = \max\{f(1), 1\} + \min\{f(1), 1\} = f(1) + 1,$$

a contradiction. Thus, no such function exists.

Hungary-Israel Mathematical Competition, 2001 (Individual)

Find all continuous functions $f : \mathbb{R} \to \mathbb{R}$ such that, for all real x,
$$f(f(x)) = f(x) + x.$$

Solution

Note that $f(x) = \varphi x$ and $f(x) = -x/\varphi$ are solutions of the problem, where $\varphi = (1 + \sqrt{5})/2$ is the golden ratio. We will prove that there is no other solution.

Let f be a solution. The function f must be injective, because if $f(a) = f(b)$, then $a = f(f(a)) - f(a) = f(f(b)) - f(b) = b$. Thus, f is a bijection from $\mathbb{R}$ onto $f(\mathbb{R})$. Since f is continuous, it follows that f is strictly monotonic on $\mathbb{R}$. Moreover, $f(f(0)) = f(0)$; whence, $f(0) = 0$.

Since f is monotonic, it follows that $\lim\limits_{x \to +\infty} f(x)$ is either a real number or $\pm\infty$. If $\lim\limits_{x \to +\infty} f(x) = L \in \mathbb{R}$, then, using the continuity of f, we obtain
$$f(L) - L = \lim_{x \to +\infty} \bigl(f(f(x)) - f(x)\bigr) = \lim_{x \to +\infty} x = +\infty,$$
which is absurd. We have a similar result for $\lim\limits_{x \to -\infty} f(x)$. Therefore, f is unbounded above and unbounded below, and $f(\mathbb{R}) = \mathbb{R}$.

Let f^0 denote the identity function on $\mathbb{R}$. For each positive integer n, let $f^{n+1} = f \circ f^n$ and $f^{-(n+1)} = f^{-1} \circ f^{-n}$. Since we are given that $f^2 = f^1 + f^0$, it follows that $f^{n+2} = f^{n+1} + f^n$ for all $n \in \mathbb{Z}$. Solving this difference equation, we get
$$f^n = \frac{g}{\sqrt{5}} \varphi^n + \frac{h}{\sqrt{5}} \left(-\frac{1}{\varphi}\right)^n,$$
where $g = \frac{1}{\varphi} f^0 + f^1$ and $h = \varphi f^0 - f^1$; that is, $g(x) = \frac{1}{\varphi} x + f(x)$ and $h(x) = \varphi x - f(x)$. Note that for all $x \in \mathbb{R}$, since $\varphi > 1$, we have
$$\lim_{n \to -\infty} \frac{g(x)}{\sqrt{5}} \varphi^n = 0 \quad \text{and} \quad \lim_{n \to +\infty} \frac{h(x)}{\sqrt{5}} \left(-\frac{1}{\varphi}\right)^n = 0.$$

Case 1. f is increasing.

We will show that $h(x) = 0$ for all $x \in \mathbb{R}$. Then we will have $f(x) = \varphi x$ for all $x \in \mathbb{R}$.

First consider any $x > 0$. Since $f(0) = 0$ and f is increasing, we deduce that $f^n(x) > 0$ for all $n \in \mathbb{Z}$. If $h(x) < 0$, then
$$\lim_{n \to -\infty} f^{2n}(x) = \lim_{n \to -\infty} \frac{h(x)}{\sqrt{5}} \left(-\frac{1}{\varphi}\right)^{2n} = -\infty,$$
which is impossible, since $f^n(x) > 0$ for all $n \in \mathbb{Z}$. Similarly, if $h(x) > 0$, then $\lim\limits_{n \to -\infty} f^{2n+1}(x) = -\infty$, and again we have a contradiction. Therefore, $h(x) = 0$.

Now consider any $x < 0$. In this case, we have $f^n(x) < 0$ for all $n \in \mathbb{Z}$. In much the same manner as above, we see that if $h(x) \neq 0$, then either $\lim_{n \to -\infty} f^{2n}(x) = \infty$ or $\lim_{n \to -\infty} f^{2n+1}(x) = \infty$, giving a contradiction. Thus, $h(x) = 0$.

Finally, we note that $h(0) = 0 - f(0) = 0$.

Case 2. f is decreasing.

By an argument similar to the argument in Case 1, but with $x \to +\infty$ instead of $x \to -\infty$, we see that $g(x) = 0$ for all $x \in \mathbb{R}$. Then $f(x) = -x/\varphi$ for all $x \in \mathbb{R}$.

Hungary-Israel Mathematical Competition, 2001 (Individual)

Let $P(x) = x^3 - 3x + 1$. Find the polynomial Q whose roots are the fifth power of the roots of P.

Solution

Let a, b, c be the roots of P. Then $a+b+c = 0$, $ab+bc+ca = -3$, and $abc = -1$. The required polynomial Q is given by

$$\begin{aligned} Q(x) &= (x-a^5)(x-b^5)(x-c^5) \\ &= x^3 - (a^5+b^5+c^5)x^2 + (a^5b^5+b^5c^5+c^5a^5)x - a^5b^5c^5 \\ &= x^3 - S_5 x^2 + T_5 x + 1, \end{aligned}$$

where $S_5 = a^5 + b^5 + c^5$ and $T_5 = a^5 b^5 + b^5 c^5 + c^5 a^5$.

Generalizing our definition of S_5, we define $S_n = a^n + b^n + c^n$, for each positive integer n. Observe that $T_5 = \frac{1}{2}(S_5^2 - S_{10})$. Therefore, the determination of $Q(x)$ will be complete if we can calculate S_5 and S_{10}.

We have $S_1 = a+b+c = 0$ and $S_2 = (a+b+c)^2 - 2(ab+bc+ca) = 6$. Since a, b, and c are roots of P, they each satisfy the equation $x^3 = 3x - 1$. It follows that $S_{n+3} = 3S_{n+1} - S_n$ for all $n \geq 0$. Therefore,

$$\begin{aligned} S_3 &= 3 \times 0 - 3 = -3 \\ S_4 &= 3 \times 6 - 0 = 18 \\ S_5 &= 3 \times (-3) - 6 = -15 \\ S_6 &= 3 \times 18 - (-3) = 57 \\ S_7 &= 3 \times (-15) - 18 = -63 \\ S_8 &= 3 \times 57 - (-15) = 186 \\ S_{10} &= 3 \times 186 - (-63) = 621. \end{aligned}$$

Now we have $T_5 = \frac{1}{2}(S_5^2 - S_{10}) = \frac{1}{2}(15^2 - 621) = -198$. Thus, $Q(x) = x^3 + 15x^2 - 198x + 1$.

2nd Hong Kong Mathematical Olympiad, 1999

Determine all functions $f : \mathbb{R} \to \mathbb{R}$ such that, for all $x, y \in \mathbb{R}$,

$$f(x + yf(x)) = f(x) + xf(y).$$

Solution

The solutions are the zero function and the identity function. Clearly, these functions satisfy the given condition (which we will denote by $\mathcal{C}$). Conversely, we show that any non-zero function f satisfying $\mathcal{C}$ is given by $f(x) = x$ for all real x. The proof goes through eleven steps.

(1) $f(0) = 0$.
 This follows from $\mathcal{C}$ with $x = 1$ and $y = 0$.

(2) $f(x) = 0 \Longrightarrow x = 0$.
 If $f(x) = 0$, then $0 = f(x) = f(x + yf(x)) = xf(y)$, which implies that $x = 0$, since we can choose y such that $f(y) \neq 0$.

(3) $f(-1) = -1$.
 This follows from step (2) and $\mathcal{C}$ with $x = y = -1$.

(4) $f(x - f(x)) = f(x) - x$ for all x.
 Apply $\mathcal{C}$ with $y = -1$ and use (3).

(5) $f(1) = 1$.
 Applying $\mathcal{C}$ with $x = 1 - f(1)$ and $y = 1$, and using (4), we have $f(0) = -(1 - f(1))^2$; then use (1).

(6) $f(t - 1) = f(t) - 1$ for all t.
 Apply $\mathcal{C}$ with $x = 1$ and $y = t - 1$, and use (5).

(7) $f(u) = u \Longrightarrow f(tu) = uf(t)$ for all t.
 Suppose that $f(u) = u$. Then

$$f(tu) = f(u + (t - 1)f(u)) = f(u) + uf(t - 1),$$

using $\mathcal{C}$ with $x = u$ and $y = t - 1$. Using (6), we get

$$f(tu) = f(u) + u(f(t) - 1) = uf(t).$$

(8) f is odd.
 Apply (7) with $u = -1$, noting (3).

(9) $f(v) = -v \Longrightarrow f(tv) = -vf(t)$ for all t.
 This is analogous to (7). Apply $\mathcal{C}$ with $x = v$ and $y = 1 - t$, and use (8) and (6) to get $f(1 - t) = -f(t - 1) = 1 - f(t)$.

(10) $f(x^2) = (f(x))^2$ for all x.

If $f(x) = x$, the result follows from (7) with $u = t = x$. Otherwise, let $v = x - f(x)$. Then $v \neq 0$ and by (4) we have $f(v) = -v$. Now, (9) and $\mathcal{C}$ with $y = \frac{x}{v}$ together yield

$$\begin{aligned} f(x^2) &= f(vx + xf(x)) = -vf\left(x + \frac{x}{v}f(x)\right) \\ &= -vf(x) + x(-v)f\left(\frac{x}{v}\right) = -vf(x) + xf(x) = (f(x))^2. \end{aligned}$$

(11) $f(r) \geq 0$ if $r \geq 0$.
    Use $x = \sqrt{r}$ in (10).

Now, let x be any real number. If $x \geq f(x)$, then $f(x - f(x)) \geq 0$ by (11); hence, by (4), we have $f(x) \geq x$, and eventually $f(x) = x$. If $x \leq f(x)$, then $f(f(x) - x) \geq 0$ and, since f is odd, we have $-f(x - f(x)) \geq 0$. Therefore, $x - f(x) \geq 0$ and eventually $f(x) = x$. In either case, $f(x) = x$.

17th Balkan Mathematical Olympiad, 2000

Find all the functions $f : \mathbb{R} \to \mathbb{R}$ with the property that
$$f(xf(x) + f(y)) = (f(x))^2 + y,$$
for any real numbers x and y.

Solution

The functions $f(x) = x$ and $f(x) = -x$ are solutions. We claim that these are the only solutions.

Let $f : \mathbb{R} \to \mathbb{R}$ such that, for all real numbers x and y,
$$f(xf(x) + f(y)) = (f(x))^2 + y. \tag{1}$$
Let $f(0) = a$. Setting $x = 0$ in (1) yields
$$f(f(y)) = a^2 + y \tag{2}$$
for all $y \in \mathbb{R}$. This equation shows that f is a bijection. As a consequence, there exists b such that $f(b) = 0$. Setting $x = b$ in (1), we get, for all $y \in \mathbb{R}$,
$$f(f(y)) = y. \tag{3}$$
Comparing (2) and (3), we see that $a = 0$ (and hence $b = 0$). Then, substituting $y = 0$ into (1), we get
$$f(xf(x)) = (f(x))^2, \tag{4}$$
for all $x \in \mathbb{R}$. Now, setting $x = f(t)$ in (4) gives
$$f(tf(t)) = t^2, \tag{5}$$
for all $t \in \mathbb{R}$. Comparing (4) and (5), we get
$$(f(x))^2 = x^2 \tag{6}$$
for all $x \in \mathbb{R}$. Thus, for each $x \in \mathbb{R}$, we have either $f(x) = x$ or $f(x) = -x$.

Suppose there exist non-zero numbers α and β such that $f(\alpha) = -\alpha$ and $f(\beta) = \beta$. Then, taking $x = \alpha$ and $y = \beta$ in (1), we get
$$f(-\alpha^2 + \beta) = \alpha^2 + \beta,$$
which contradicts (6). We conclude that $f(x) = x$ for all $x \in \mathbb{R}$ or $f(x) = -x$ for all $x \in \mathbb{R}$.

49th Mathematical Olympiad of Lithuania, 2000

A function $f : \mathbb{R} \to \mathbb{R}$ satisfies the following equation for all real x and y:

$$(x+y)(f(x) - f(y)) = f(x^2) - f(y^2).$$

Find: (a) one such function; (b) all such functions.

Solution

Any affine function $x \mapsto ax + b$ (for some real numbers a, b) clearly satisfies the functional equation. Conversely, we show that any solution is an affine function.

Let f be any solution. Set $b = f(0)$ and $g(x) = f(x) - b$ ($x \in \mathbb{R}$). It is readily seen that g is a solution as well and satisfies $g(0) = 0$. Taking $y = 0$ in the given equation (written for g) yields

$$(x+0)(g(x) - g(0)) = g(x^2) - g(0);$$

that is, $xg(x) = g(x^2)$ for all real x. Substituting $-x$ for x gives us $-xg(-x) = g(x^2)$, and it follows that g is an odd function. Thus,

$$(x+y)(g(x) - g(y)) = g(x^2) - g(y^2) = (x-y)(g(x) + g(y)).$$

Thus, $xg(y) = yg(x)$ for all x and y. As a result, $g(x)/x$ is constant on $\mathbb{R} \setminus \{0\}$ and $g(x) = xg(1)$ for all real numbers x. It follows that $f(x) = xg(1) + b$ and f is an affine function.

XXXVI Spanish Mathematical Olympiad, 2000

Let $P(x) = x^4 + ax^3 + bx^2 + cx + 1$ and $Q(x) = x^4 + cx^3 + bx^2 + ax + 1$, with a, b, c real numbers and $a \neq c$. Find conditions on a, b, and c so that $P(x)$ and $Q(x)$ have two common roots. In this case, solve the equations $P(x) = 0$, $Q(x) = 0$.

Solution

The common roots of $P(x)$ and $Q(x)$ are among the roots of the polynomial
$$P(x) - Q(x) = (a-c)x(x^2-1).$$
Thus, they are among -1, 0, and 1. Since $P(0) = Q(0) = 1$, the common roots must be -1 and 1. Substituting these values of x into the equations $P(x) = 0$ and $Q(x) = 0$, we get the conditions $a + b + c + 2 = 0$ and $a - b + c - 2 = 0$, from which we get $b = -2$ and $a + c = 0$.

When these conditions are satisfied, $P(x)$ and $Q(x)$ can be written as
$$P(x) = x^4 + ax^3 - 2x^2 - ax + 1 = (x^2 - 1)(x^2 + ax - 1),$$
$$Q(x) = x^4 - ax^3 - 2x^2 + ax + 1 = (x^2 - 1)(x^2 - ax - 1).$$

Thus, the roots of $P(x) = 0$ are
$$-1, \quad 1, \quad \frac{-a - \sqrt{a^2+4}}{2}, \quad \text{and} \quad \frac{-a + \sqrt{a^2+4}}{2},$$
and the roots of $Q(x) = 0$ are
$$-1, \quad 1, \quad \frac{a - \sqrt{a^2+4}}{2}, \quad \text{and} \quad \frac{a + \sqrt{a^2+4}}{2}.$$

XXXVI Spanish Mathematical Olympiad, 2000

Show that there is no function $f : \mathbb{N} \to \mathbb{N}$ such that $f(f(n)) = n + 1$.

Solution

Suppose for the purpose of contradiction that such a function exists. Let m be the natural number defined by $f(0) = m$. Then $f(k) = m + k$ for $k = 0$. Let $k \geq 0$ be any natural number such that $f(k) = m + k$. Then $f(m+k) = f(f(k)) = k+1$. Thus, $f(k+1) = f(f(m+k)) = m+k+1$. It follows by induction that $f(k) = m+k$ for all $k \in \mathbb{N}$. However, this yields $f(m) = 2m$, while $f(0) = m$ yields $f(m) = f(f(0)) = 1$, a contradiction.

Note: Bornsztein points out that the well-known problem #4 of the 1987 IMO was to prove that there is no function $f : \mathbb{N} \to \mathbb{N}$ such that $f(f(n)) = n + 1987$. Any solution of this problem contains a remark that the result holds just because 1987 is odd.

8th Macedonian Mathematical Olympiad

Does there exist a function $f : \mathbb{N} \to \mathbb{N}$ such that for every $n \geq 2$,
$$f(f(n-1)) = f(n+1) - f(n)?$$

Solution

We will prove that there is no such function.

Assume, for a contradiction, that f is a function such that
$$f(f(n-1)) = f(n+1) - f(n). \tag{1}$$

Then, for all $n \geq 2$, we have $f(n+1) - f(n) = f(f(n-1)) > 0$. Thus, f is increasing on $\{2, 3, 4, \ldots\}$. Since $f(2) \geq 1$, it follows that $f(n) \geq n-1$ for all $n \geq 2$.

If $f(n) = n - 1$ for all $n \geq 2$, then, for $n \geq 4$,
$$f(f(n-1)) = f(n-2) = n-3$$
$$\text{and} \quad f(n+1) - f(n) = n - (n-1) = 1.$$

This contradicts (1) for $n \geq 5$.

Hence, there exists $n_0 \geq 2$ such that $f(n_0) \geq n_0$. As above, we deduce that $f(n) \geq n$ for all $n \geq n_0$.

Repeating the same reasoning twice (once for n and once for $n+1$), we prove that there exists $a \geq 2$ such that $f(n) \geq n+2$ for all $n \geq a$.

Now, let $b = f(a)$. Then $b - 2 \geq a$ and
$$f(f(a)) = f(a+2) - f(a+1),$$
$$f(f(a+1)) = f(a+3) - f(a+2),$$
$$\vdots \qquad \vdots$$
$$f(f(b-2)) = f(b) - f(b-1).$$

Summing, we obtain
$$f(f(a)) + f(f(a+1)) + \cdots + f(f(b-2))$$
$$= f(b) - f(a+1) = f(f(a)) - f(a+1).$$

Thus,
$$0 \leq f(f(a+1)) + \cdots + f(f(b-2)) = -f(a+1) < 0,$$

a contradiction.

13th Irish Mathematical Olympiad

Let $f(x) = 5x^{13} + 13x^5 + 9ax$. Find the least positive integer a such that 65 divides $f(x)$ for every integer x.

Solution

Note that $f(x) = x(5x^{12} + 13x^4 + 9a)$ and $65 = 5 \times 13$. Let x be an integer.

If $x \equiv 0 \pmod{13}$, then $f(x) \equiv 0 \pmod{13}$. If $x \not\equiv 0 \pmod{13}$, then $5x^{12} + 13x^4 + 9a \equiv 5 + 9a \pmod{13}$, using Fermat's Little Theorem, and hence,

$$f(x) \equiv 0 \pmod{13} \quad \text{if and only if} \quad a \equiv -2 \pmod{13}. \qquad (1)$$

If $x \equiv 0 \pmod 5$, then $f(x) \equiv 0 \pmod 5$. If $x \not\equiv 0 \pmod 5$, then $5x^{12} + 13x^4 + 9a \equiv 3 + 9a \pmod 5$, using Fermat's Little Theorem again, and hence,

$$f(x) \equiv 0 \pmod 5 \quad \text{if and only if} \quad a \equiv -2 \pmod 5. \qquad (2)$$

From (1) and (2), we deduce that the least positive integer a such that $f(x) \equiv 0 \pmod{65}$ for all integers x, is defined by $a + 2 = \text{lcm}(5, 13) = 65$. Thus, the desired integer is $a = 63$.

13th Irish Mathematical Olympiad

10 Let $p(x) = a_0 + a_1 x + \cdots + a_n x^n$ be a polynomial with non-negative real coefficients. Suppose that $p(4) = 2$ and $p(16) = 8$. Prove that $p(8) \leq 4$, and find, with proof, all such polynomials with $p(8) = 4$.

Solution

Applying the Cauchy-Schwarz Inequality,

$$(u_0 v_0 + u_1 v_1 + \cdots + u_n v_n)^2 \leq (u_0^2 + u_1^2 + \cdots + u_n^2)(v_0^2 + v_1^2 + \cdots + v_n^2),$$

with $u_i = \sqrt{a_i}\, 2^i$ and $v_i = \sqrt{a_i}\, 4^i$ for $1 \leq i \leq n$, we get

$$p(8)^2 \leq p(4) p(16) = 2 \cdot 8 = 16.$$

Taking the square root gives $p(8) \leq 4$ (since $p(8) \geq 0$).

If equality holds, then $v_i = c u_i$ for some real c and for all i. Then, since $v_i = 2^i u_i$ for all i, all u_is but one must be equal to zero. This implies that $p(x) = a_i x^i$ for some i. Then it is easy to see that $i = 1$ and $a_i = \frac{1}{2}$, so that $p(x) = \frac{1}{2} x$. Conversely, if $p(x) = \frac{1}{2} x$, then equality holds.

Singapore Mathematical Olympiad, 2002 (Open Section, Part A)

Let $f(x)$ be a function which satisfies

$$f(29 + x) = f(29 - x),$$

for all values of x. If $f(x)$ has exactly three real roots α, β, and γ, determine the value of $\alpha + \beta + \gamma$.

Solution

Since f has exactly three real roots and f has the same value at points symmetric about 29, one of the roots must be 29. Let $\gamma = 29$. The other two roots, α and β, must be symmetric about 29; hence, $\alpha = 29 + x$ amd $\beta = 29 - x$ for some real number $x \neq 0$. Therefore,

$$\alpha + \beta + \gamma = (29 + x) + (29 - x) + 29 = 87.$$

Singapore Mathematical Olympiad, 2002 (Open Section, Part A)

It is given that the polynomial $p(x) = x^3 + ax^2 + bx + c$ has three distinct positive integer roots and $p(2002) = 2001$. Let $q(x) = x^2 - 2x + 2002$. It is also given that the polynomial $p(q(x))$ has no real roots. Determine the value of a.

Solution

The polynomial $q(x) = (x - 1)^2 + 2001$ takes on every value in the interval $[2001, \infty)$. Since $p(q(x))$ has no real roots, all three roots of $p(x)$ must be less than 2001. Denoting the roots of $p(x)$ by x_1, x_2, x_3, we have

$$p(x) = (x - x_1)(x - x_2)(x - x_3).$$

Then $p(2002) = (2002 - x_1)(2002 - x_2)(2002 - x_3) = 2001 = 3 \cdot 23 \cdot 29$. Since each factor $2002 - x_i$ is a positive integer, we must have

$$\{2002 - x_1,\ 2002 - x_2,\ 2002 - x_3\} = \{3,\ 23,\ 29\},$$

and hence $\{x_1, x_2, x_3\} = \{1999, 1979, 1973\}$. Then

$$a = -(x_1 + x_2 + x_3) = -5951.$$

Singapore Mathematical Olympiad, 2002 (Open Section, Part B) — 4

Find all real-valued functions $f : \mathbb{Q} \longrightarrow \mathbb{R}$ defined on the set of all rational numbers $\mathbb{Q}$ satisfying the conditions

$$f(x+y) = f(x) + f(y) + 2xy,$$

for all x, y in $\mathbb{Q}$ and $f(1) = 2002$. Justify your answers.

Solution

It is readily checked that the function $r \mapsto r(r+2001)$ is a solution. We will show that it is unique. Let f be an arbitrary solution. Denote by $\mathcal{C}$ the given condition $f(x+y) = f(x) + f(y) + 2xy$, and fix $z \in \mathbb{Q}$, $z > 0$. It is easily proved that $f(nz) = n(f(z) + (n-1)z^2)$ for all $n \in \mathbb{N}$ (by induction, using $\mathcal{C}$ with $x = nz$ and $y = z$ for the inductive step). Then, for all $n \in \mathbb{N}$,

$$2002 = f(1) = f\left(n \times \frac{1}{n}\right) = n\left(f\left(\frac{1}{n}\right) + (n-1)\cdot\frac{1}{n^2}\right),$$

and thus, $f\left(\frac{1}{n}\right) = \frac{1}{n}\left(2001 + \frac{1}{n}\right)$. Then, for all positive integers m and n,

$$f\left(\frac{m}{n}\right) = f\left(m \times \frac{1}{n}\right) = m\left(f\left(\frac{1}{n}\right) + (m-1)\cdot\frac{1}{n^2}\right) = \frac{m}{n}\left(\frac{m}{n} + 2001\right).$$

Thus, $f(r) = r(r+2001)$ holds for all positive $r \in \mathbb{Q}$. Since $f(0) = 0$ (condition $\mathcal{C}$ with $x = y = 0$) and $f(-r) = 2r^2 - f(r)$ (condition $\mathcal{C}$ with $x = r$ and $y = -r$), it can be verified that $f(r) = r(r+2001)$ actually holds for all $r \in \mathbb{Q}$. This completes the proof.

15th Korean Mathematical Olympiad

Find all functions $f : \mathbb{R} \to \mathbb{R}$ satisfying $f(x - y) = f(x) + xy + f(y)$ for every $x \in \mathbb{R}$ and every $y \in \{f(x) \mid x \in \mathbb{R}\}$, where $\mathbb{R}$ is the set of all real numbers.

Solution

The functions $\theta : x \mapsto 0$ and $\phi : x \mapsto -x^2/2$ are the solutions for f.

It is readily checked that θ and ϕ are solutions. We show that there is no other solution.

Let f be a function such that
$$f(x - y) = f(x) + xy + f(y), \tag{1}$$
for every $x \in \mathbb{R}$ and $y \in \{f(x) \mid x \in \mathbb{R}\}$. Let $a = f(0)$. Taking $x = 0$ in (1), we get
$$f(-y) = a + f(y). \tag{2}$$
Next we take $x = y$ in (1) to get
$$f(y) = \tfrac{1}{2}(a - y^2). \tag{3}$$
Equations (2) and (3) hold whenever $y \in f(\mathbb{R}) = \{f(x) \mid x \in \mathbb{R}\}$.

For $x \in \mathbb{R}$ and $y \in f(\mathbb{R})$, we have $xy - f(x - y) = -f(y) - f(x)$ (from (1)), and therefore,
$$f(xy - f(x - y)) = f(-f(y) - f(x)). \tag{4}$$
Using (1) and (3), the left side L of (4) may be rewritten as
$$\begin{aligned}
L &= f(xy) + xyf(x - y) + \tfrac{1}{2}a - \tfrac{1}{2}(f(x - y))^2 \\
&= f(xy) + \tfrac{1}{2}a + \tfrac{1}{2}f(x - y)(2xy - f(x - y)) \\
&= f(xy) + \tfrac{1}{2}a + \tfrac{1}{2}(xy + f(x) + f(y))(xy - f(x) - f(y)) \\
&= f(xy) + \tfrac{1}{2}a + \tfrac{1}{2}x^2y^2 - \tfrac{1}{2}(f(x))^2 - \tfrac{1}{2}(f(y))^2 - f(x)f(y).
\end{aligned}$$
Using (1), (2), and (3), the right side R of (4) may be rewritten as
$$\begin{aligned}
R &= f(-f(y)) - f(x)f(y) + f(f(x)) \\
&= a + f(f(y)) - f(x)f(y) + f(f(x)) \\
&= a + \tfrac{1}{2}a - \tfrac{1}{2}(f(y))^2 - f(x)f(y) + \tfrac{1}{2}a - \tfrac{1}{2}(f(x))^2 \\
&= 2a - \tfrac{1}{2}(f(y))^2 - f(x)f(y) - \tfrac{1}{2}(f(x))^2.
\end{aligned}$$
From $L = R$, we deduce that $f(xy) + \tfrac{1}{2}x^2y^2 = \tfrac{3}{2}a$. With $x = 0$, we obtain $a = \tfrac{3}{2}a$; hence, $a = 0$. Thus, $f(xy) = -\tfrac{1}{2}x^2y^2$. If $f \neq \theta$, there exists $b \neq 0$ with $b \in f(\mathbb{R})$, and then, for all $r \in \mathbb{R}$,
$$f(r) = f\left(b \cdot \frac{r}{b}\right) = -\frac{1}{2} \cdot b^2 \cdot \frac{r^2}{b^2} = -\frac{1}{2} \cdot r^2.$$
Therefore, $f = \phi$, and the proof is complete.

15th Korean Mathematical Olympiad

For $n \geq 3$, let $S = a_1 + a_2 + \cdots + a_n$ and $T = b_1 b_2 \cdots b_n$ for positive real numbers $a_1, a_2, \ldots, a_n, b_1, b_2, \ldots, b_n$, where the numbers b_i are pairwise distinct.

(a) Find the number of distinct real zeroes of the polynomial

$$f(x) = (x - b_1)(x - b_2) \cdots (x - b_n) \sum_{j=1}^{n} \frac{a_j}{x - b_j}.$$

(b) Prove the inequality

$$\frac{1}{n-1} \sum_{j=1}^{n} \left(1 - \frac{a_j}{S}\right) b_j > \left(\frac{T}{S} \sum_{j=1}^{n} \frac{a_j}{b_j}\right)^{\frac{1}{n-1}}.$$

Solution

(a) Without loss of generality, suppose that $b_1 < b_2 < \cdots < b_n$. Let

$$R(x) = \sum_{j=1}^{n} \frac{a_j}{x - b_j} = \frac{f(x)}{(x - b_1)(x - b_2) \cdots (x - b_n)}. \quad (1)$$

For each $j \in \{1, 2, \ldots, n\}$, the function $x \mapsto a_j/(x - b_j)$ is strictly decreasing on the intervals $(-\infty, b_j)$ and (b_j, ∞). Therefore, R is strictly decreasing on each of the intervals $(-\infty, b_1), (b_1, b_2), \ldots, (b_{n-1}, b_n)$, and (b_n, ∞). Furthermore, for each $j \in \{1, 2, \ldots, n\}$, since

$$\lim_{x \to b_j^-} \frac{a_j}{x - b_j} = -\infty \quad \text{and} \quad \lim_{x \to b_j^+} \frac{a_j}{x - b_j} = \infty,$$

we have $\lim_{x \to b_j^-} R(x) = -\infty$ and $\lim_{x \to b_j^+} R(x) = \infty$. Therefore, R vanishes exactly once on each interval (b_j, b_{j+1}). Since $R(x) < 0$ for $x \in (-\infty, b_1)$ and $R(x) > 0$ for $x \in (b_n, \infty)$, it follows that R has exactly $n - 1$ real zeroes (and these are distinct).

From (1), we see that

$$f(x) = \sum_{j=1}^{n} a_j p_j(x), \quad (2)$$

where $p_j(x) = \prod_{i \neq j} (x - b_i)$. Thus, $f(x)$ is a polynomial of degree $n - 1$ and has at most $n - 1$ real zeroes. Each of the $n - 1$ distinct real zeroes of R is also a zero of $f(x)$ (as we see from (1)). We conclude that $f(x)$ has exactly $n - 1$ distinct real zeroes.

(b) Let $r_1, r_2, \ldots, r_{n-1}$ be the zeroes of $f(x)$. From (2), we see that the coefficient of x^{n-1} in $f(x)$ is S. Therefore,

$$f(x) = S(x - r_1)(x - r_2) \cdots (x - r_{n-1}). \quad (3)$$

Comparing the coefficients of x^{n-2} in (2) and (3), we obtain

$$-S(r_1 + r_2 + \cdots + r_{n-1}) = \sum_{j=1}^{n} a_j \left(-\sum_{i \neq j} b_i\right),$$

and hence,

$$r_1 + r_2 + \cdots + r_{n-1} = \frac{1}{S} \sum_{j=1}^{n} a_j \sum_{i \neq j} b_i = \frac{1}{S} \sum_{i=1}^{n} b_i \sum_{j \neq i} a_j$$

$$= \sum_{i=1}^{n} b_i \left(1 - \frac{a_i}{S}\right).$$

Applying the AM–GM Inequality, we get

$$\frac{1}{n-1} \sum_{i=1}^{n} b_i \left(1 - \frac{a_i}{S}\right) = \frac{r_1 + r_2 + \cdots + r_{n-1}}{n-1}$$

$$> (r_1 r_2 \cdots r_{n-1})^{\frac{1}{n-1}}. \qquad (4)$$

(The inequality is strict because the r_j are not all equal.)

Setting $x = 0$ in (2) and (3) gives

$$\sum_{j=1}^{n} a_j \prod_{i \neq j} (-b_i) = S(-1)^{n-1} r_1 r_2 \cdots r_{n-1};$$

that is, $r_1 r_2 \cdots r_{n-1} = \frac{T}{S} \sum_{j=1}^{n} \frac{a_j}{b_j}$. This, together with (4), gives the desired inequality.

15th Irish Mathematical Olympiad, (First Paper)

Denote by $\mathbb{Q}$ the set of rational numbers. Determine all functions $f : \mathbb{Q} \to \mathbb{Q}$ such that
$$f(x + f(y)) = y + f(x), \quad \text{for all } x, y \in \mathbb{Q}.$$

Solution

Setting $x = 0$ in the given condition, we find that $f(f(y)) = f(0) + y$ for all $y \in \mathbb{Q}$. If $f(x) = f(y)$ for some $x, y \in \mathbb{Q}$, then $f(f(x)) = f(f(y))$; hence, $f(0) + x = f(0) + y$, from which we get $x = y$. Thus, f is injective.

Setting $y = 0$ in the given condition, we obtain $f(x + f(0)) = f(x)$ for all $x \in \mathbb{Q}$. Since f is injective, we get $x = x + f(0)$, and thus, $f(0) = 0$. Then $f(f(y)) = y$ for all $y \in \mathbb{Q}$.

Now, for all $x, y \in \mathbb{Q}$,
$$f(x + y) = f(x + f(f(y))) = f(x) + f(y).$$

Hence, by an easy induction, $f(nx) = nf(x)$ for all $n \in \mathbb{N}^*$. Now let x be any positive rational number. Setting $x = r/s$ with $r, s \in \mathbb{N}^*$, we have $sf(x) = f(sx) = f(r) = rf(1)$; hence, $f(x) = (r/s)f(1) = xf(1)$.

Now we note that, for all $x \in \mathbb{Q}$,
$$0 = f(0) = f(x - x) = f(x) + f(-x),$$

and therefore, $f(-x) = -f(x) = -xf(1)$. Thus, $f(x) = xf(1)$ for all $x \in \mathbb{Q}$. Moreover, by setting $x = f(1)$, we find that $f(f(1)) = f(1)f(1)$. From above, we have $f(f(1)) = 1$. Thus, $f(1) = \pm 1$. We conclude that either $f(x) = x$ for all $x \in \mathbb{Q}$ or $f(x) = -x$ for all $x \in \mathbb{Q}$.

It is now easy to check that $f(x) = x$ and $f(x) = -x$ are solutions.

Vietnamese Mathematical Olympiad, 2003

Find all polynomials $P(x)$ with real coefficients, satisfying the relation

$$(x^3 + 3x^2 + 3x + 2)P(x-1) = (x^3 - 3x^2 + 3x - 2)P(x)$$

for every real number x.

Solution

Let P be a real polynomial satisfying the given condition; that is,

$$(x+2)(x^2+x+1)P(x-1) = (x-2)(x^2-x+1)P(x). \qquad (1)$$

Since the polynomials $x+2$, x^2+x+1, $x-2$, and x^2-x+1 are irreducible over $\mathbb{R}$, we see that $P(x)$ is divisible by $(x+2)(x^2+x+1)$. Thus,

$$P(x) = (x+2)(x^2+x+1)Q(x) \qquad (2)$$

for some real polynomial $Q(x)$, and (1) yields

$$P(x-1) = (x-2)(x^2-x+1)Q(x). \qquad (3)$$

Taking $x = 2$ in (3) gives $P(1) = 0$, which implies that $P(x)$ is divisible by $x - 1$. Since $P(-2) = 0$ (in view of (2)), taking $x = -1$ in (3) gives $Q(-1) = 0$. Therefore, $Q(x)$ is divisible by $x + 1$, and then so is $P(x)$. Lastly, taking $x = 1$ in (1) gives $P(0) = 0$, so that $P(x)$ is divisible by x as well. Summing up, we see that $P(x) = (x+2)(x+1)x(x-1)(x^2+x+1)S(x)$ for some real polynomial $S(x)$.

Now, substituting into (1), we obtain $S(x) = S(x-1)$ for all $x \in \mathbb{R}$, and an immediate induction shows that $S(n) = S(0)$ for all non-negative integers n. Thus, the polynomial $S(x) - S(0)$ has infinitely many roots. It follows that $S(x)$ is a constant polynomial.

Conversely, substituting $P(x) = k(x+2)(x+1)x(x-1)(x^2+x+1)$ in (1) leads to an identity, for all $k \in \mathbb{R}$.

In conclusion, the solutions of the problem are the polynomials of the form $P(x) = k(x+2)(x+1)x(x-1)(x^2+x+1)$, where $k \in \mathbb{R}$.

Vietnamese Mathematical Olympiad, 2003

Let $P(x) = 4x^3 - 2x^2 - 15x + 9$ and $Q(x) = 12x^3 + 6x^2 - 7x + 1$.

(i) Prove that each of these polynomials has three distinct real roots.

(ii) Let α and β be the greatest roots of $P(x)$ and $Q(x)$, respectively. Prove that $\alpha^2 + 3\beta^2 = 4$.

Solution

(i) The polynomial P is a continuous function, and it is easily checked that $P(-2) < 0$, $P(-\frac{15}{8}) > 0$, $P(0) > 0$, $P(1) < 0$, and $P(\frac{15}{8}) > 0$. Hence, P has three distinct roots α, α_1, and α_2 satisfying

$$\alpha \in \left(1, \tfrac{15}{8}\right), \quad \alpha_1 \in (0, 1), \quad \alpha_2 \in \left(-2, -\tfrac{15}{8}\right). \tag{1}$$

Similarly, Q has three distinct roots, β, β_1, and β_2 such that

$$\beta \in \left(\tfrac{1}{3}, 1\right), \quad \beta_1 \in \left(0, \tfrac{1}{3}\right), \quad \beta_2 \in (-2, -1). \tag{2}$$

(ii) A polynomial $S(x)$ whose roots are exactly α^2, α_1^2, and α_2^2 is readily obtained by taking S such that $S(x^2) = -P(x) \cdot P(-x)$. Here, since

$$\begin{aligned}S(x^2) &= -(9 - 2x^2 + x(4x^2 - 15))(9 - 2x^2 - x(4x^2 - 15)) \\ &= x^2(4x^2 - 15)^2 - (9 - 2x^2)^2,\end{aligned}$$

we obtain $S(x) = 16x^3 - 124x^2 + 261x - 81$. Similarly, the roots of the polynomial

$$T(x) = 144x^3 - 204x^2 + 37x - 1$$

are β^2, β_1^2, and β_2^2.

Now, transforming $T(x)$ through the relation $y = 4 - 3x$ (that is, substituting $x = (4-y)/3$ in $T(x)$) leads to $T((4-y)/3) = -\frac{1}{3}S(y)$, which shows that $\{\alpha^2, \alpha_1^2, \alpha_2^2\} = \{4 - 3\beta^2, 4 - 3\beta_1^2, 4 - 3\beta_2^2\}$. Furthermore, from (1) and (2),

$$\alpha^2 \in \left(1, \tfrac{225}{64}\right), \quad \alpha_1^2 \in (0, 1), \quad \alpha_2^2 \in \left(\tfrac{225}{64}, 4\right),$$

so that $\alpha_1^2 < \alpha^2 < \alpha_2^2$ and

$$4 - 3\beta^2 \in \left(1, \tfrac{11}{3}\right), \quad 4 - 3\beta_1^2 \in \left(\tfrac{11}{3}, 4\right), \quad 4 - 3\beta_2^2 \in (-8, 1),$$

so that $4 - 3\beta_2^2 < 4 - 3\beta^2 < 4 - 3\beta_1^2$. The desired result, $\alpha^2 = 4 - 3\beta^2$, follows.

Vietnamese Mathematical Olympiad, 2003

Let f be a function defined on the set of real numbers $\mathbb{R}$, taking values in $\mathbb{R}$, and satisfying the condition $f(\cot x) = \sin 2x + \cos 2x$ for every x belonging to the open interval $(0, \pi)$. Find the least and the greatest values of the function $g(x) = f(x) \cdot f(1 - x)$ on the closed interval $[-1, 1]$.

Solution

We show that g has a minimum value of $4 - \sqrt{34}$ and a maximum value of $1/25$ on $[-1, 1]$.

For $x \in (0, \pi)$, we have

$$\begin{aligned} f(\cot x) &= \sin 2x + \cos 2x = 2 \sin x \cos x + \cos^2 x - \sin^2 x \\ &= \sin^2 x (2 \cot x + \cot^2 x - 1) = \frac{\cot^2 x + 2 \cot x - 1}{\cot^2 x + 1}. \end{aligned}$$

Hence, for all $x \in \mathbb{R}$,

$$f(x) = f(\cot(\cot^{-1}(x))) = \frac{x^2 + 2x - 1}{x^2 + 1}.$$

Since $g(\frac{1}{2} + h) = f(\frac{1}{2} + h) f(\frac{1}{2} - h) = g(\frac{1}{2} - h)$ for all real h, it is sufficient to study the values of $g(\frac{1}{2} + h)$ for $h \in [0, \frac{3}{2}]$. An easy computation gives

$$g\left(\frac{1}{2} + h\right) = \frac{16h^4 - 136h^2 + 1}{16h^4 + 24h^2 + 25} = 1 - 8\phi(h^2),$$

where ϕ is defined on $[0, \frac{9}{4}]$ by $\phi(x) = \frac{20x + 3}{16x^2 + 24x + 25}$. Since the derivative $\phi'(x)$ has the same sign as $-80x^2 - 24x + 107$, it follows that ϕ reaches its maximum on $[0, \frac{9}{4}]$ at $x_0 = \frac{\sqrt{34}}{5} - \frac{3}{20}$ with $\phi(x_0) = \frac{20}{32x_0 + 24} = \frac{\sqrt{34} - 3}{8}$ and its minimum at 0 with $\phi(0) = \frac{3}{25}$. Thus, the extreme values of $g(\frac{1}{2} + h)$ are $1 - 8\phi(x_0) = 4 - \sqrt{34}$ (minimum) and $1 - 8\phi(0) = \frac{1}{25}$ (maximum).

8th Macedonian Mathematical Competition

Does there exist a function $f : \mathbb{N} \to \mathbb{N}$ such that for every $n \geq 2$,
$$f(f(n-1)) = f(n+1) - f(n)?$$

Solution

Suppose that f is such a function. Then, for all $n \geq 2$,
$$f(n+1) - f(n) = f(f(n-1)) \geq 1.$$
Thus, f is increasing for $n \geq 2$. By induction, we have $f(n) \geq n-1$ for all $n \geq 2$. If $f(8) \leq 9$, then
$$3 = 9 - 6 \geq f(8) - f(7) = f(f(6)) \geq f(5) \geq 4,$$
which is a contradiction. Hence, $f(8) \geq 10$. Then
$$f(10) > f(10) - f(9) = f(f(8)) \geq f(10),$$
which is again a contradiction.

British Mathematical Competition, 2003 (Round II)

Let f be a function from the set of non-negative integers into itself such that, for all $n \geq 0$,

(i) $(f(2n+1))^2 - (f(2n))^2 = 6f(n) + 1$, and

(ii) $f(2n) \geq f(n)$.

How many numbers less than 2003 are there in the image of f?

Solution

Since $6f(n)+1$ is odd, both $f(2n+1) - f(2n)$ and $f(2n+1) + f(2n)$ must be odd, by (i). If $f(2n+1) - f(2n) \geq 3$, then

$$f(2n+1) + f(2n) \leq \tfrac{1}{3}(6f(n)+1) = 2f(n) + \tfrac{1}{3}$$

and

$$2f(2n) = (f(2n+1)+f(2n)) - (f(2n+1)-f(2n)) \leq 2f(2n) + \tfrac{1}{3} - 3,$$

which is a contradiction. Therefore, $f(2n+1) - f(2n) = 1$, and (i) reduces to $f(2n+1) + f(2n) = 6f(n) + 1$. Solving the system, we get $f(2n) = 3f(n)$ and $f(2n+1) = 3f(n) + 1$ for all $n \geq 0$.

Now we show, by induction on n, that if the base–2 representation of n is $a_k 2^k + \cdots + a_1 2 + a_0$, then $f(n) = a_k 3^k + \cdots + a_1 3 + a_0$. Since $f(0) = 0$, the claim is true for $n = 0$. Assume that $n \geq 1$ and the claim is true up to $n-1$. Write $n = a_k 2^k + \cdots + a_1 2 + a_0$. Then

$$\begin{aligned} f(n) &= f(2(a_k 2^{k-1} + \cdots + a_1) + a_0) \\ &= 3f(a_k 2^{k-1} + \cdots + a_1) + a_0 = a_k 3^k + \cdots + a_1 3 + a_0, \end{aligned}$$

where the last step is by the induction hypothesis. Finally, since $3^7 > 2003$, it follows that $3^6 + \cdots + 3 + 1 = 1093$ is the largest number less than 2003 in the image of f. So there are $2^7 = 128$ numbers less than 2003 in the image of f.

Ukrainian Mathematical Olympiad, (11 Form)

Find all functions $f : \mathbb{R} \to \mathbb{R}$ such that
$$f(xf(x) + f(y)) = x^2 + y$$
for all $x \in \mathbb{R}$ and $y \in \mathbb{R}$.

Solution

It is easy to verify that both $f(x) = x$ and $f(x) = -x$ satisfy the functional equation. We will prove that these are the only solutions.

Let f be a solution. Setting $x = 0$ in the functional equation gives $f(f(y)) = y$ for all $y \in \mathbb{R}$. Hence, for all $x, y \in \mathbb{R}$,

$$x^2 + y = f(xf(x) + f(y)) = f\bigl(f(x)f(f(x)) + f(y)\bigr) = f(x)^2 + y.$$

Thus, $f(x)^2 = x^2$ for all $x \in \mathbb{R}$. If $f(1) = 1$, then

$$\begin{aligned} 1 + 2x + x^2 &= f(1+x)^2 = f\bigl(1f(1) + f(f(x))\bigr)^2 \\ &= \bigl(1^2 + f(x)\bigr)^2 = 1 + 2f(x) + x^2, \end{aligned}$$

and thus $f(x) = x$ for all $x \in \mathbb{R}$. Similarly, if $f(1) = -1$, then

$$\begin{aligned} 1 - 2x + x^2 &= f(-1+x)^2 = f\bigl(1f(1) + f(f(x))\bigr)^2 \\ &= \bigl(1^2 + f(x)\bigr)^2 = 1 + 2f(x) + x^2, \end{aligned}$$

and thus $f(x) = -x$ for all $x \in \mathbb{R}$.

Belarus Mathematical Olympiad, 2003

Find all functions f from the real numbers to the real numbers such that, for any real numbers x and y,

$$f(xy)(f(x) - f(y)) = (x - y)f(x)f(y).$$

Solution

Clearly, the zero function is a solution.
We will show that the non-zero solutions are the functions $\Phi_{a,K}$ defined by

$$\Phi_{a,K}(x) = \begin{cases} ax & \text{if } x \in K, \\ 0 & \text{if } x \notin K, \end{cases}$$

where K is a subgroup of the multiplicative group $\mathbb{R}^* = \mathbb{R} \setminus \{0\}$ and $a \in \mathbb{R}^*$.

Consider $a \in \mathbb{R}^*$ and a subgroup K of $\mathbb{R}^*$. We show that $f = \Phi_{a,K}$ satisfies

$$f(xy)(f(x) - f(y)) = (x - y)f(x)f(y) \qquad (1)$$

for all $x, y \in \mathbb{R}$.

If $x, y \in K$, then $xy \in K$ and (1) holds since it can be rewritten as $axy(ax - ay) = (x - y)ax \cdot ay$.

If $x, y \notin K$, then $\Phi(x) = \Phi(y) = 0$ and (1) is true.

If, say, $x \in K$, $y \notin K$, certainly $xy \notin K$ (otherwise $y = xy \cdot \frac{1}{x}$ would be in K). Thus, $\Phi(xy) = \Phi(y) = 0$ and (1) holds.

Conversely, let f be any function from $\mathbb{R}$ to $\mathbb{R}$ satisfying (1), and assume that f is not the zero function. Then $f(x_0) \neq 0$ for some $x_0 \in \mathbb{R}$. Taking $x = 1$ and $y = 0$ in (1) yields $f(0) = 0$ (hence, $x_0 \neq 0$). Then $y = 1$ gives

$$f(x)(f(x) - ax) = 0, \qquad (2)$$

where we set $a = f(1)$. This relation (2) with $x = x_0$ shows that $a \in \mathbb{R}^*$ and, more generally, that $f(x) = ax$ if $f(x) \neq 0$.

Now, let

$$K = \{x \in \mathbb{R}^* \mid f(x) = ax\} = \{x \in \mathbb{R} \mid f(x) \neq 0\}.$$

Clearly $1 \in K$. If $x_1, x_2 \in K$ with $x_1 \neq x_2$, then $f(x_1) = ax_1$ and $f(x_2) = ax_2$. Then (1) with $x = x_1$ and $y = x_2$ gives $f(x_1 x_2) = ax_1 x_2$; hence, $x_1 x_2 \in K$. Similarly, (1) with $x = x_1$ and $y = 1/x_1$ shows that $1/x_1 \in K$ and, lastly, (1) with $x = x_1^2$ and $y = 1/x_1$ gives $x_1^2 \in K$. We have proved that K is a subgroup of $\mathbb{R}^*$ and that $f = \Phi_{a,K}$. This completes the proof.

Indian TST for IMO, 2003

Find all functions $f : \mathbb{R} \to \mathbb{R}$ such that, for all x, y in $\mathbb{R}$, we have

$$f(x+y) + f(x)f(y) = f(x) + f(y) + f(xy). \qquad (1)$$

Solution

Clearly, $f \equiv 0$, $f \equiv 2$, and $f(x) = x$ are solutions. We show that they are the only solutions.

Setting $x = 0 = y$ in (1), we get $(f(0))^2 = 2f(0)$. If $f(0) = 2$, then, by letting $y = 0$ in (1), we see that $f(x) = 2$ for all $x \in \mathbb{R}$; that is, $f \equiv 2$.

Suppose therefore that $f(0) = 0$. Let $a = f(1)$. Setting $x = 1$ and $y = -1$, we get $af(-1) = a + 2f(-1)$; that is, $f(-1) = a/(a-2)$. Now successively substituting $(x-1, 1)$, $(-x+1, -1)$, and $(-x, 1)$ for (x, y) in (1), we get

$$f(x) + (a-2)f(x-1) = a, \qquad (2)$$

$$f(-x) + \frac{2}{a-2}f(-x+1) - f(x-1) = \frac{a}{a-2}, \qquad (3)$$

$$f(-x+1) + (a-2)f(-x) = a. \qquad (4)$$

Eliminating $f(x-1)$ and $f(-x+1)$ in (2), (3), and (4) gives

$$f(x) - (a-2)f(-x) = 0. \qquad (5)$$

Replacing x by $-x$ in (5), we obtain

$$f(-x) - (a-2)f(x) = 0. \qquad (6)$$

If $a \notin \{1, 3\}$, then by eliminating $f(-x)$ in (5) and (6), we see that $f(x) = 0$ for all $x \in \mathbb{R}$; that is, $f \equiv 0$.

If $a = 3$, then (2) gives $f(x) = 3 - f(x-1)$ for all $x \in \mathbb{R}$. Hence, $f(2) = 3 - f(1) = 0$ and $f(\frac{5}{2}) = 3 - f(\frac{3}{2}) = f(\frac{1}{2})$. On the other hand, by substituting $(2, \frac{1}{2})$ for (x, y) in (1), we get $f(\frac{5}{2}) = f(\frac{1}{2}) + f(1) = f(\frac{1}{2}) + 3$, a contradiction.

Finally, consider $a = 1$. Then (2) gives $f(x) = f(x-1) + 1$ for all $x \in \mathbb{R}$. By induction, $f(x+n) = f(x) + n$, for all $x \in \mathbb{R}$ and $n \in \mathbb{Z}$. In particular, $f(n) = f(0+n) = f(0) + n = n$ for all $n \in \mathbb{Z}$. Substituting n for y in (1), we obtain $nf(x) = f(nx)$ for all $x \in \mathbb{R}$ and $n \in \mathbb{Z}$. Hence, if $r = m/n \in \mathbb{Q}$ and $x \in \mathbb{R}$, then

$$mf(x) = f(mx) = f(n \cdot rx) = nf(rx);$$

that is, $f(rx) = rf(x)$. In particular, $f(r) = f(r \cdot 1) = rf(1) = r$. Setting $y = r$ in (1) gives $f(x+r) = f(x) + r$. Also, setting $y = x$ in (1), we get $(f(x))^2 = f(x^2)$. Thus, $f(x) \geq 0$ if $x \geq 0$. Since $f(-x) = -f(x)$, we have $f(x) \leq 0$ if $x \leq 0$.

Now, let $x \in \mathbb{R}$ be fixed. If $r \in \mathbb{Q}$ and $r < x$, then

$$f(x) = f(x-r+r) = f(x-r) + r \geq r,$$

since $f(x-r) \geq 0$. It follows that $f(x) \geq x$. Similarly, $f(x) \leq r$ for all $r \in \mathbb{Q}$ such that $r > x$, which implies that $f(x) \leq x$. Thus, $f(x) = x$.

Indian TST for IMO, 2003

Let $P(x)$ be a polynomial with integer coefficients such that $P(n) > n$ for all positive integers n. Suppose that for each positive integer m, there is a term in the sequence $P(1)$, $P(P(1))$, $P(P(P(1)))$, ... which is divisible by m. Show that $P(x) = x + 1$.

Solution

We will use the notation $P^{(i)}(1)$ for the ith term in the sequence $P(1)$, $P(P(1))$, ... The condition $P(n) > n$ implies that $\deg(P) \geq 1$ and the leading coefficient of P is positive.

If $P(x) = x + b$, then $1 + b = P(1) > 1$, which implies that $b \geq 1$. It is easy to see that $P(x) = x + 1$ satisfies all the conditions. If $b \geq 2$, then $P(1) \equiv 1 \pmod{b}$ and, by induction, $P^{(i)}(1) \equiv 1 \pmod{b}$ for all $i \geq 1$.

If $P(x) = 2x + b$, then $2 + b = P(1) > 1$, which implies that $b \geq 0$. If $b = 0$, then, by induction, $P^{(i)}(1) = 2^i$ for all $i \geq 1$.

We consider together all the remaining cases: (i) $P(x) = 2x + b$ with $b \geq 1$; (ii) $P(x) = ax + b$ with $a \geq 3$; and (iii) $\deg(P) \geq 2$. In these cases, there exists $N \in \mathbb{N}$ such that $P(n) > 2n$ for all $n \geq N$. Since $1 < P(1) < P(P(1)) < P(P(P(1))) < \cdots$, there exists $k \in \mathbb{N}$ such that $P^{(k)}(1) \geq N$. Let $r = P^{(k)}(1)$ and $m = P^{(k+1)}(1) - P^{(k)}(1)$. Then $r \geq N$ and $m = P(r) - r > r$. For $1 \leq i \leq k$, we have $1 < P^{(i)}(1) \leq r < m$; thus, m does not divide any $P^{(i)}(1)$ for $1 \leq i \leq k$. Moreover, note that $P^{(k+1)}(1) = m + r \equiv r \pmod{m}$. Assume as an induction hypothesis that $P^{(i)}(1) \equiv r \pmod{m}$ for some $i \geq k + 1$. Then

$$\begin{aligned} P^{(i+1)}(1) &= P(P^{(i)}(1)) \equiv P(r) = P(P^{(k)}(1)) \\ &= P^{(k+1)}(1) \equiv r \pmod{m} . \end{aligned}$$

Hence, $P^{(i)}(1) \equiv r \pmod{m}$ for all $i \geq k + 1$.

Singapore Mathematical Olympiad, 2003

Find all real-valued functions $f : \mathbb{Q} \longrightarrow \mathbb{R}$ defined on the set of all rational numbers $\mathbb{Q}$ satisfying the conditions

$$f(x+y) = f(x) + f(y) + 2xy,$$

for all x, y in $\mathbb{Q}$ and $f(1) = 2002$. Justify your answers.

Solution

Let $g(x) = f(x) - x^2$ for $x \in \mathbb{Q}$. Then, for all $x, y \in \mathbb{Q}$,

$$\begin{aligned} g(x+y) &= f(x+y) - (x+y)^2 \\ &= f(x) + f(y) + 2xy - (x+y)^2 = g(x) + g(y). \end{aligned}$$

This is the well-known Cauchy Equation, whose solutions are $g(x) = cx$, where c is a constant. Hence, $f(x) = x^2 + 2001x$ (since $f(1) = 2002$).

Iranian Mathematical Olympiad, 2002 (First Round)

Let δ be a symbol such that $\delta \neq 0$ and $\delta^2 = 0$. Define
$$\mathbb{R}[\delta] = \{a + b\delta \mid a, b \in \mathbb{R}\}$$
$$a + b\delta = c + d\delta \iff a = c \text{ and } b = d,$$
$$(a + b\delta) + (c + d\delta) = (a + c) + (b + d)\delta,$$
$$(a + b\delta) \cdot (c + d\delta) = ac + (ad + bc)\delta.$$

Let $P(x)$ be a polynomial with real coefficients. Show that $P(x)$ has a multiple root in $\mathbb{R}$ if and only if $P(x)$ has a non-real root in $\mathbb{R}[\delta]$.

Solution

Let $a, b \in \mathbb{R}$. An easy induction shows that for all $n \in \mathbb{N}$, we have $(a + b\delta)^n = a^n + na^{n-1}b\delta$. It follows that
$$P(x + y\delta) = P(x) + P'(x)y\delta \tag{1}$$
for all real x and y.

If $P(x)$ has a multiple root x_0 in $\mathbb{R}$, then $P(x_0) = P'(x_0) = 0$ and, from (1), we have $P(x_0 + \delta) = 0$. Thus, $P(x)$ has a non-real root in $\mathbb{R}[\delta]$.

Conversely, if $P(a + b\delta) = 0$ for some real numbers a and b with $b \neq 0$, then, from (1) again, $P(a) + P'(a)b\delta = 0 = 0 + 0\delta$. Hence, $P(a) = P'(a) = 0$ and a is a multiple real root of $P(x)$.

Belarusian Mathematical Olympiad, 2002 (Category A)

Let
$$P(x) = (x+1)^p(x-3)^q = x^n + a_1 x^{n-1} + a_2 x^{n-2} + \cdots + a_{n-1} x + a_n,$$
where p and q are positive integers.

(a) Given that $a_1 = a_2$, prove that $3n$ is a perfect square.

(b) Prove that there exist infinitely many pairs (p, q) of positive integers p and q such that the equality $a_1 = a_2$ is valid for the polynomial $P(x)$.

Solution

(a) The roots of $P(x)$ are evidently -1 and 3, with respective multiplicities p and q (where $p + q = n$). Denote these roots by $r_1, r_2, \ldots, r_n$, where $r_1 = r_2 = \cdots = r_p = -1$ and $r_{p+1} = r_{p+2} = \cdots = r_n = 3$.

Using the well-known relations between roots and coefficients, we have $\sum_{i=1}^{n} r_i = -a_1$ and $\sum_{1 \leq i < j \leq n} r_i r_j = a_2$. On the other hand, using the known values of the roots, we obtain $\sum_{i=1}^{n} r_i = -p + 3q$ and

$$\sum_{1 \leq i < j \leq n} r_i r_j = \binom{p}{2}(-1)^2 + pq(-1)(3) + \binom{q}{2}(3)^2$$
$$= \tfrac{1}{2}p(p-1) - 3pq + \tfrac{9}{2}q(q-1).$$

Thus,
$$a_1 = p - 3q \quad \text{and} \quad a_2 = \tfrac{1}{2}p(p-1) - 3pq + \tfrac{9}{2}q(q-1).$$

Therefore, $a_1 = a_2$ if and only if
$$p^2 - 3p - 6pq + 9q^2 - 3q = 0. \tag{1}$$

From (1), we deduce that p is divisible by 3, which in turn forces q to be a multiple of 3. Let $p = 3a$ and $q = 3b$.

Thus, $a_1 = a_2$ if and only if $a^2 - a - 6ab + 9b^2 - b = 0$, which is equivalent to $(a - 3b)^2 = a + b$. It follows that $a_1 = a_2$ if and only if $3n = 9(a + b) = 9(a - 3b)^2$, and we are done.

(b) From above, $a_1 = a_2$ if and only if $(a - 3b)^2 = a + b$, where $p = 3a$ and $q = 3b$. Let $a - 3b = x$. Then $a_1 = a_2$ if and only if

$$(a, b) = \left(\tfrac{3}{4}x(x+1), \tfrac{1}{4}x(x-1)\right). \tag{2}$$

Since a and b are positive integers, we choose any positive integer t and, letting $x = 4t$ in (2), we deduce that, for $(p, q) = (36t^2 + 3t, 12t^2 - 3t)$, the polynomial $P(x)$ satisfies $a_1 = a_2$.

Belarusian Mathematical Olympiad, 2002 (Category A)

Does there exist a surjective function $f : \mathbb{R} \to \mathbb{R}$ such that the expression $f(x+y) - f(x) - f(y)$ takes exactly two values 0 and 1 for various real x and y?

Solution

Yes. For example, let

$$f(x) = \begin{cases} x & \text{if } x \leq 0 \\ x - 1 & \text{if } x > 0. \end{cases}$$

Then f is clearly surjective. Now take any $x \leq y$.

(i) If $x \leq y \leq 0$, then $f(x+y) = x+y = f(x) + f(y)$.
(ii) If $x \leq 0 < y$ and $x+y \leq 0$, then $f(x+y) = x+y = f(x) + f(y) + 1$.
(iii) If $x \leq 0 < y$ and $x+y > 0$, then $f(x+y) = x+y-1 = f(x) + f(y)$.
(iv) If $0 < x \leq y$, then $f(x+y) = x+y-1 = f(x) + f(y) + 1$.

www.ingramcontent.com/pod-product-compliance
Lightning Source LLC
Chambersburg PA
CBHW080515090426
42734CB00015B/3056